BRAIN STORY

BRAIN STORY

UNLOCKING OUR INNER WORLD OF EMOTIONS, MEMORIES, IDEAS AND DESIRES

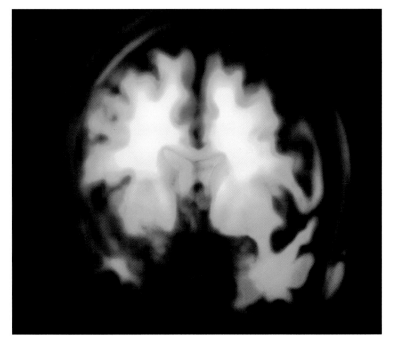

SUSAN GREENFIELD

DK

London, New York, Sydney, Delhi, Paris,
Munich, and Johannesburg

First published in 2000 by BBC Worldwide Limited,
Woodlands, 80 Wood Lane, London W12 0TT

Executive producer: Alan Bookbinder
Series producer: Sam Roberts
Producers: Andrew Cohen, Zoe Heron, Sam Roberts,
Thecla Schreuders, Sam Starbuck
Commissioning editor: Sheila Ableman
Project editor: Martha Caute
Text editor: Ben Morgan
Designer: Linda Blakemore
Picture researcher: Frances Abraham

This book is published to accompany the television series
Brain Story, first broadcast on BBC 2 in 2000

First US edition published in 2001 by
Dorling Kindersley Publishing, Inc.
95 Madison Avenue
New York, New York 10016

Publisher: Sean Moore
Editorial Director: Chuck Wills
Project Editor: Barbara Minton
Art Director: Dirk Kaufman
Production Director: David Proffit

ISBN 0-7894-7839-0

Set in Sabon and GillSans by BBC Worldwide
Printed and Bound in France by Imprimerie Pollina s.a. - n° L81692
Color separations by Imprimerie Pollina s.a.

ACKNOWLEDGMENTS

I would like to thank the following for their help on this project: at BBC Television,
Glenwyn Benson, Alan Bookbinder, Jill Buckland, Andrew Cohen, Milla Harrison, Zoe
Heron, Gina Kocjancic, John Lynch, Anna Mishcon, Basia Pietluch, Sam Roberts, Thecla
Schreuders, Sam Starbuck, Rami Tzabar, Jackie Williamson; at BBC Worldwide, Frances
Abraham, Sheila Ableman, Linda Blakemore, Martha Caute, Ben Morgan, Shirley Patton;
at the University of Oxford, Dr Ole Paulsen (Pharmacology) and Professor Nicholas
Rawlins (Experimental Psychology). SG

CONTENTS

For Peter

PREFACE

It seems amazing now, in this first year of the new century, that the 1990s were designated 'The Decade of the Brain' – as if the huge mysteries of our most fascinating organ could be solved within ten years! Then again, perhaps it was not so short sighted. The previous century saw enormous strides in our understanding of the generic brain and how it functions at the nuts-and-bolts level. Of course, there is still a long way to go, but many of the basic concepts are now familiar to brain researchers: for many, it is just a question of filling in the gaps, while adhering to paradigms laid down earlier. But for others of us, the adventure is just beginning.

Now we have a crude grasp of a few of the words, we wish to master the complete language – the grammar of the brain as well as its vocabulary. In short, this century is the time when it is ripe to look beyond isolated brain phenomena – the release of certain chemicals, the lighting up of certain brain regions in a scan – and to try and make sense of how the brain might be working as a whole, how it might account for the differences between a human and a chimp, or between any two individual people, and how it might generate emotions and consciousness. Personally, I'm happy to shut the door on the Decade of the Brain, and welcome instead, the Century of the Mind.

It was with these sentiments that we planned the BBC 2 series *Brain Story*. We were clear from the outset that this was to be no detailed, pedagogic-type exercise of the type that I might have used for my medical students. The viewers, we decided, were not about to take an exam in neuroscience, and therefore did not need detailed or exhaustive facts about everything that was known so far about the brain. Instead, they would probably be more interested to learn about those aspects of brain research that touched directly on their personal lives, and indeed posed the most exciting and penetrating questions about the human condition that not just a scientist, but anyone can ask.

Our goal was to seek out worldwide the most paradigm-shifting research that would throw new light on the big questions. Inevitably, therefore, we were to encounter controversy, as different researchers had different views, claims and findings. We tried to preserve this dynamism – always a feature of ongoing research – and to present as balanced a picture as possible. But since we are dealing with material that is far from cut and dried, it is inevitable that there is a personal element to some of the science we are describing. The personal element also comes in with the type of studies that we wished to feature. Much of what we learn about brain function is derived from personal cases of strokes, accidents, and other mishaps. We were deeply indebted therefore to the patients who agreed to tell their own stories in front of the camera. It is these personal elements, both from the stances of scientists and patients, that I have tried to capture in this book. To the best of my knowledge other books on the brain (including my own) have never been quite so involved with the nature of the individual.

I was particularly thrilled at the prospect of a book to accompany the series since it enabled me to expand on much of what could not be included, for time reasons and other practical constraints, in the programmes themselves. I have been able, freed up from the demands of working with a TV crew, to include far more facts, case stories, and reflections than would ever have been possible on screen. It is the flavour, not the literal content, of the series that has been the driving force: to bring together the different threads of what we are learning about the brain that makes you, you. As such, this book falls midway between a simple lay guide to neuroscience and a more philosophical text. It is a personal slant on the mind, but one grounded in science: it is the story of *your* brain.

Susan Greenfield, Oxford, February 2000

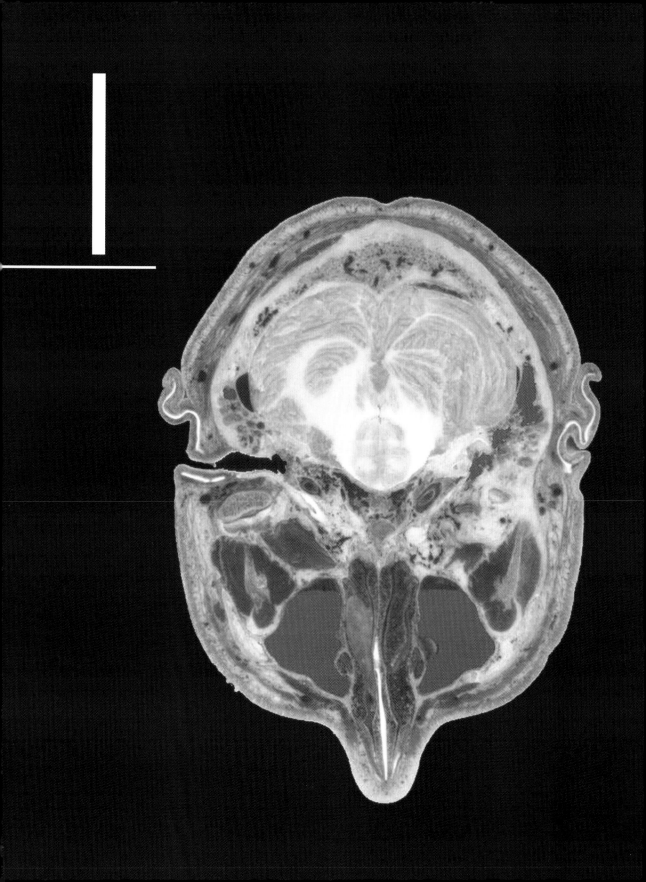

IN SEARCH OF THE SELF

Sarah was lying on her side, eyes shut. Slowly she started to stir, her body making restless, uncoordinated movements as she drifted back into consciousness. Her mouth was dry and she asked for a drink. Julie wiped her mouth with a medicated swab – it would be dangerous to allow someone to take in fluids during surgery. Julie was an anaesthetist; Sarah was lying not in bed but on an operating table. I was watching, for the first time in my life, brain surgery on a conscious patient.

Sarah's problem was that she had a brain tumour. Fortunately, it was benign and easily accessible, lying on the outer surface of the brain (the cortex). The neurosurgeon, my friend Henry, intended to remove the growth by suction, using a device called an aspirator, which looks like a hollow pen. The difficulty was that the tumour had grown near to the part of the cortex related to speech. Henry knew that a serious risk in the procedure he was about to perform was that, although the tumour would be sucked out, the aspirator might also suck out key bits of Sarah's brain and leave her with severe speech problems. One way around the difficulty was for Henry to be absolutely sure that his pen went nowhere near the part of the brain that Sarah used for speech. So, first he had to find out where exactly in Sarah's brain the key areas lay. He could find these areas by placing electrodes on the surface of Sarah's brain and applying minute electric currents. The currents would jam the normal processes so that, for the brief period of stimulation, Sarah would be unable to speak. Of course, for the procedure to work, Sarah would have to be speaking – and that would not be possible if she was deeply anaesthetized.

It might seem macabre that Sarah could be fully conscious while Henry was poking around inside her head in this way. But the brain itself does not have any pain sensors, so once the painful part of the surgery had been completed – removing a part of Sarah's skull – she could be woken up. After a perfectly normal conversation about her forthcoming wedding, I strolled to the other side of the plastic curtain that was draped across Sarah's side, dividing her front from her back. Here, Henry was at

◀◀ A section through the human head seen from above. The intricately folded cerebellum at the back of the brain can be seen at the upper centre as a pale brown symmetrical structure.

work, peering down a microscope poised above a 6-inch-wide circular opening in Sarah's head. Incredibly, the blotchy red stuff inside the hole was what I had just been talking to, and it was still chatting to Julie. Henry let me look down the operating microscope and I saw an almost liquid, textured sea of tiny blood vessels marbling a creamy, quivering mass. This was all there was to Sarah, or indeed to any of us. Here was the very reality that has fascinated me all my adult life: that we are but sludgy brains, and that, somehow, a character and a mind are generated in this soupy mess.

Sarah had declined the option of watching the operation on a monitor. Nonetheless, it would have been possible for her to observe her own brain – the thing on the TV would have been seeing itself. My own brain turned cartwheels at the thought of a brain watching itself. Meanwhile, Sarah went on counting as Henry applied the electric current. Suddenly, her voice wavered and slurred, and then, when the current was turned off, came back loud and clear, running effortlessly through the numbers. Fortunately, the part of her brain that was making such a valuable contribution to her speech was far from the deep red tumour. Henry applied his aspirator. 'Don't suck too much away,' Sarah joked, as slurpy sounds began to emanate from behind the curtain. Soon the duty pathologist, who was testing the brain tissue for signs of tumour, gave the all clear. As Julie infused some anaesthetic into Sarah's arm, she drifted back to sleep, to wake later, fully recovered and skull complete.

I was left in a state of awe. Leaving aside Sarah's incredible courage and sense of humour, Julie's compassion and skill, not to mention Henry's dexterity and knowledge, I had been faced with the ultimate banality of all that we hold dear, our very individuality – and it was nothing but a creamy, quivering mass of some sort of *stuff*. Yet, somehow, that stuff contained secrets, thoughts and feelings.

The same stark truth can be seen in stories with a less happy outcome than Sarah's. For 30 years, Dick Lingham was a music teacher. He played regularly at school recitals and church events across his home county of Cambridgeshire. But eight years ago, his wife Lynn began to notice changes in Dick that she couldn't quite explain.

'It was very subtle things at first, he seemed to be losing his sense of humour, and he was vague and subdued. I noticed that his driving had become hazardous: he'd pull out into the path of oncoming traffic without even thinking.'

By 1993, the subtle changes had become more obvious. Although an accomplished musician, Dick suddenly found it difficult to read music

Even very simple animals, such as the sea slug *Aplysia*, have a primitive sort of brain (although *Aplysia*'s brain contains a mere 17,000 brain cells, a tiny fraction of the trillion or so found in the human brain). Animals need a brain to help them move around – the brain takes in information from the senses about the changing environment, and then coordinates appropriate responses. But animals that don't move around (and plants, for that matter) have no need of brains. The sea squirt, for instance, has a brain only during the larval stage of its life cycle, when it swims. Once it becomes an adult, it settles down in one spot and lives by filtering tiny particles of food out of sea water. Its brain, no longer needed, is consumed.

Brains come in different shapes and sizes, but a basic ground plan has prevailed throughout evolution. Every brain consists of a mass of cells called neurons (nerve cells), which form complex, interconnected networks. These networks, in turn, build up into characteristic regions that are often identifiable to the naked eye. All mammals share the three brain regions that feature in this chapter: the cerebellum, a cauliflower-shaped structure at the back that looks like an independent brain in its own right; the basal ganglia, a series of interconnected regions in the centre of the brain; and the cortex, the brain's outer layer.

The cortex will dominate much of what follows in this book. One reason why neuroscientists and psychologists are so preoccupied with this part of the brain is that, in humans, it is enormous. As primates evolved and their brains became larger, the cortex grew out of all proportion and its surface area expanded. To fit into the confines of the skull, the cortex had to fold up, forming distinct lobes separated by deep fissures. The folded pattern gives the human brain its characteristic wrinkly appearance. Less intelligent mammals, such as rats and mice, have a completely smooth cortex. More advanced animals, such as cats, have a slightly wrinkly cortex. But primates – and especially humans – have such a wrinkly cortex that their brains resemble walnuts.

and play the piano simultaneously. Tasks that had once seemed easy became an effort. He was sent to Addenbrooke's Hospital in Cambridge for tests, and the results showed he was suffering from Pick's disease, a condition in which the front part of the brain slowly disintegrates, causing dementia – loss of memory and increasing confusion. In spite of the damage to the brain that takes place in Pick's disease, a small number of patients develop surprising new talents. Dick discovered an artistic ability that he never knew he had, and he now paints compulsively for up to four hours a day.

If a diseased brain can express new abilities while it is being destroyed, what does that tell us about a normal brain? As we shall see later in this book, scientists are discovering a great deal about the physical and chemical processes that take place in the brain. But we find it very difficult to relate these processes to the complex thoughts, feelings and abilities that make up our lives. Dick's 'creative dementia' illustrates the problem – different skills and talents, such as painting or playing the

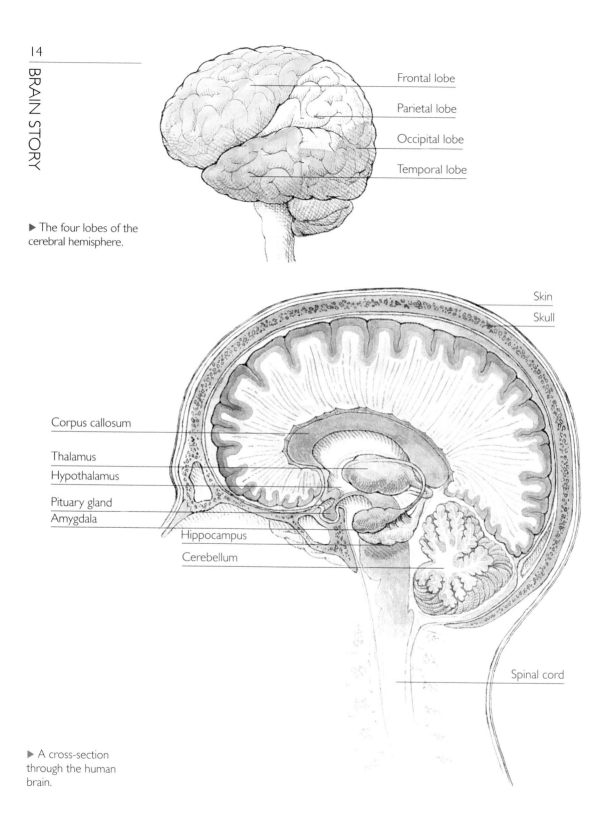

Frontal lobe

Parietal lobe

Occipital lobe

Temporal lobe

▶ The four lobes of the cerebral hemisphere.

Skin

Skull

Corpus callosum

Thalamus

Hypothalamus

Pituary gland

Amygdala

Hippocampus

Cerebellum

Spinal cord

▶ A cross-section through the human brain.

piano, are not simply housed away in discrete, autonomous parts of the brain. Somehow, the brain is functioning holistically, configuring and reconfiguring as we grow, age or succumb to disease. Sadly, for Dick, his new artistic tendencies will soon disappear along with everything else – a tragic testament to the stark truth that we are nothing but brains.

Lynn again: 'In many ways I feel I've lost my husband – he is not the same person. He's changed from a caring, kind man into someone who needs to be looked after continually. He can be nasty and crude; he can't even cross the road himself, and doesn't care if there is a car coming. I just want my husband back'.

'I just want my wife back again.' The man in front of me was speaking in a matter-of-fact tone. He had just made a 200-mile journey in the hope that my lecture on Parkinson's disease might give some clue, some new insight, into his wife's condition. Before her illness, they had travelled widely and had enjoyed a full social life. Now she had difficulty moving and was so embarrassed by her frozen facial muscles and the tremor in her hands that she refused to leave the house. Her disease was a secret that had crippled not just her body but her lifestyle, and that of her husband too.

Then there was my former neighbour, whose mother was suffering from Alzheimer's disease. All her time was spent tending to the shrivelled and silent figure huddled in the bed, a living corpse who did not recognize her own daughter, and who stared glassy eyed at a world that had long ago ceased to have meaning. Slowly, she had retraced the path she first walked as a young girl, when she learnt the names of the people and objects around her – but now she was travelling backwards, forgetting everything she had learnt. As the illness progressed, a lifetime of memories, skills, impressions, prejudices and thoughts gradually vanished. The body was left as a shell without a personality.

Parkinson's and Alzheimer's diseases are far from rare. In fact, they are becoming more common. Most of us could readily name celebrities, such as Ronald Reagan and Harold Wilson, the actor Terry-Thomas, the boxer Muhammad Ali, or the writer Iris Murdoch, who have suffered one or other of these two 'neurodegenerative' diseases. More recent victims are the actor Michael J. Fox, who has Parkinson's disease, and the one-time satirist and comic actor Dudley Moore, recently diagnosed with supranuclear palsy, a condition similar to Parkinson's in that key parts of the brain involved in movement slowly degenerate. Degeneration is just as it sounds: the inexorable and progressive death of key groups of brain cells (neurons).

In Alzheimer's disease, the problem is dementia, the erosion of the process of thought itself. By contrast, Parkinson's disease is primarily a disorder of movement: the patient wants to move but cannot translate the desire into an action. Their muscles are rigid, and their hands, incessantly shaking, take on a life of their own. But, in most cases, the intellect inside the head is intact – and desperate to get on with life.

During this new century, the frequency of neurodegenerative diseases is set to escalate to nightmare proportions. One estimate is that by 2040 there will be 7000 deaths a year from neurodegenerative diseases in the UK alone. And accompanying each death are an average of 12 years of misery – not only for the patient but also for carers, such as my neighbour and the man who was so desperate to turn back the clock to recover his wife. Currently, 1% of people older than 60 in the UK suffer from Parkinson's disease, and 5% of this age group suffer from Alzheimer's. Unless some new treatment is developed, the number of cases will rise steadily. According to another prediction, by the middle of the 21st century there may be as many as 14 million people afflicted by Alzheimer's disease in the USA – a horrific figure that takes no account of the carers whose lives are blighted indirectly.

The reason for this alarming trend is that both Alzheimer's and Parkinson's are usually disorders of older people. The chance of developing either rises dramatically with age – by age 70, for instance, a person has a 12% chance of having Alzheimer's disease and 1% chance of having Parkinson's. As we lead healthier lives, so the number of elderly people increases – in 2025, one in three of us will be over the age of 60. Advances in medicine mean that many previously fatal diseases are becoming easier to cure or manage, leaving neurodegeneration to claim the very part of us that is least understood, and yet most personal. Neurodegenerative diseases are more sinister and more frightening than heart attacks, or perhaps even cancer, because they strike at the very essence of one's humanity. They offer devastating proof that you are your brain.

Although the physician James Parkinson first documented almost 200 years ago the disease that was to bear his name, it is only since the 1960s that the cause of Parkinson's has been identified. To find it, we have to burrow far deeper into the brain than Henry travelled with his aspirator into Sarah's cortex. Deep inside the most basic part of the brain lies a special group of brain cells that are highly visible as two black, moustache-shaped bands. These unusual and conspicuous black bands give this part of the brain its name: the substantia nigra (from the Latin for 'black matter'). In Parkinson's disease, these two moustaches become much less pronounced and much paler.

Already, back in 1817, Parkinson suspected that the terrible trio of symptoms – lack of movement, tremor, and muscle rigidity – might be due to something amiss in the brain. But it was only once more was understood about how the brain works that the significance of the substantia nigra's loss of colour became appreciated. The black cells of the substantia nigra start to die and their function ceases. Normally, these key cells supply the brain with a substance called dopamine. Dopamine is one of the special chemicals that brain cells use to send signals to each other: it is a 'neurotransmitter'. We shall be looking at neurotransmitters, what they are, and how they work, more fully in the next chapter. For the time being, however, the main point is that the brain needs dopamine to control movements. When the key cells of the substantia nigra start to die off, there is insufficient dopamine, and the brain cannot control or generate movement properly.

Because dopamine is lacking in people with Parkinson's, an obvious way to treat the disease would be to replace it with fresh supplies. But dopamine molecules cannot pass through the walls of fat that divide the bloodstream from the brain, so dopamine taken as a drug would be useless. All is not lost, however, because the chemical from which dopamine is made – L-DOPA – actually can pass through the 'blood-brain barrier'. Hence, L-DOPA has been the treatment of choice for Parkinson's patients for the last three decades, and has increased the average life span after diagnosis from five years to approximately twelve. This dramatic improvement is not because L-DOPA cures the disease; rather, it alleviates the symptoms and stems the tide of diseases associated with poor mobility, such as pneumonia and disorders of the circulation. Within 25 minutes or so of taking a tablet, patients seem to undergo a miraculous change and can move with astonishing ease. But L-DOPA does not stop the cells of the substantia nigra from dying, and, in the end, no amount of L-DOPA can bring back the patient's ability to move normally.

A similar scenario occurs with Alzheimer's disease. This time, however, the damage starts in another part of the brain: the basal forebrain. This is lower down and nearer the front than the black moustaches of the substantia nigra. The cells affected in Alzheimer's disease use a different neurotransmitter, not dopamine this time but one called acetylcholine. Just as Parkinson's disease is treated by replenishing the brain's dwindling supply of dopamine, so current medication for Alzheimer's disease aims to boost the level of acetylcholine in the brain. This strategy seems

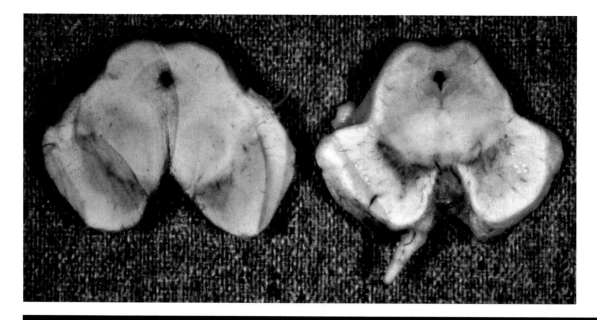

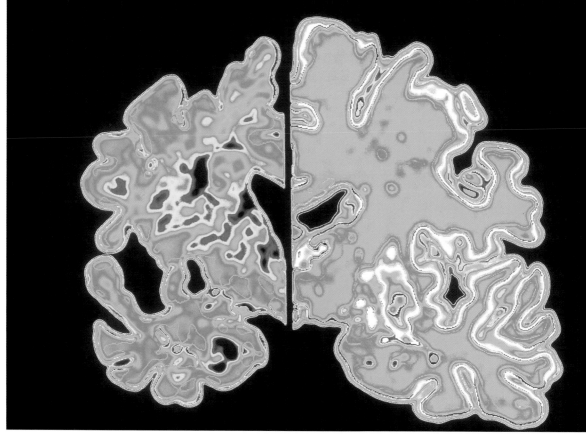

to offer some help, but the cells of the basal forebrain carry on dying and the medication cannot keep up. As a result, the parts of the brain that need acetylcholine malfunction, and this malfunction results in the all-too-familiar symptoms of dementia, such as confusion and loss of memory.

In neuroscience, one of the classic ways of investigating how the brain works has been to study people in whom an injury or disease has damaged a specific part of the brain. The idea runs that we might be able to work out the function of the affected part of the brain, or of the chemical that is missing. So, neurodegenerative disorders like Parkinson's and Alzheimer's might give us several useful lines of enquiry – not only do we know the precise regions of the brain where the problem starts, but we can also identify specific chemicals that are deficient. In Parkinson's disease, for example, does the impairment in movement that accompanies cell loss in the substantia nigra mean that the substantia nigra is the 'centre' for movement, or that dopamine is the chemical for movement?

When we look at a brain we see that it is composed of different interlocking structures of varying textures, and that it is symmetrical. A natural way of thinking about how the brain might work, therefore, has been that each brain region has a different function, and that each region is the centre for one part of our complex repertoire of behaviour. Let's see to what extent this idea might be true.

In the 19th century, a doctrine called phrenology became enormously popular. Like horoscopes today, phrenology promised people great insights into their personality and motivation – both quickly and painlessly. However, unlike horoscopes, phrenology had the semblance of science, of objective measurement. The idea, propounded originally by a German physician called Franz Gall, was that each of the bumps on the skull represented a certain character trait. These could be as clear-cut and specific as 'love of children' or 'banality'. To read your character, you simply had to survey your particular portfolio of bumps. Phrenology was soon discredited when clinical observations began to tell another story. For example, it was found that damage to various regions in the brain resulted in different types of impairment in language – and none of these regions was anywhere near the bump 'for' language described by phrenologists. More importantly, such findings indicate that more than one part of the brain is involved in complex functions such as language. It seems unlikely, therefore, that any single brain region could be the centre for language or, for that matter, the centre for any complex function.

It is still a frustrating and intriguing riddle as to how different functions relate to different parts of the brain, as the following story

◄◄▲ Sections of the midbrain showing the substantia nigra: a normal specimen (right) where the substantia nigra is clearly visible as a thin dark band on each side; a specimen showing the effects of Parkinson's disease (left) where the band is no longer apparent due to the death of the pigmental neurons.
◄◄ A slice through the brain of an Alzheimer's patient (left) which appears severely shrunken compared with that of a healthy brain (right).

illustrates. In 1990 a blood vessel burst in the brain of Isabelle Rail. A part of her brain was starved of vital oxygen and died, leaving Isabelle with severe disabilities. Yet, gradually, she recovered from all her disabilities except for one – she remained unable to 'perceive' music. Isabelle's hearing is completely intact, but her brain cannot carry out the information processing needed to understand the music she hears. For her, a piece of music evokes none of the memories or experiences that we associate with the music we love. Isabelle cannot recognize tunes that would have been familiar before her accident, although she can spot them immediately from the lyrics. And it is not just recognition that is missing. Isabelle's perception of music is so disrupted that she cannot tell whether it is in or out of tune.

What has happened in Isabelle's brain? It turns out that the damage can be tracked down to a very small area on each side of the brain, in the region of the temples. Could these tiny areas act together as the centre for music appreciation, able to function as an independent mini-brain that can process and appreciate music all on its own? A more likely scenario is that, somehow, the damaged area is a crucial component in a larger system in the brain – its loss was as disruptive as removing a sparking plug from an engine. A car will not move without a sparking plug, but that is not to say that a sparking plug can generate movement on its own. So just how might a small brain region contribute to some larger system within the brain?

Let's return to the substantia nigra, the pair of black 'moustaches' that are lost in Parkinson's disease. Far from being the centre for movement, this part of the brain turns out to be just one component – albeit an essential one – in a complex network of brain regions that all play a role in movement. (Similarly, in Alzheimer's disease, the basal forebrain is not the centre for memory or personality. Instead, it is a fountainhead for acetylcholine, and as such is a vital player in a wide-ranging network of neuron activity that ultimately emerges into the world as personality. As with the substantia nigra, the connections to other brain regions are all-important. In general, for each outward function of the brain – be it memory, language, or vision – there are many different brain regions all playing their part.)

As we have seen, a critical job for the substantia nigra is to supply other parts of the brain, particularly the striatum (a collective term for the caudate nucleus and the putamen), with the neurotransmitter dopamine. The substantia nigra and the striatum together make up part of a large and complex set of structures stretching from the top of the spinal cord to the very front of the brain, known collectively as the basal ganglia. Damage to different parts of the basal ganglia results in different impairments to movement. Degeneration in the substantia nigra, for instance, results in Parkinson's disease, characterized by a poverty of movement. Damage to the caudate nucleus results in a disease called Huntington's chorea, in which there is an excess of movement. The patient makes wild, involuntary flinging movements of the limbs that arguably resembles dancing, hence the term chorea, from the Greek for 'dance'. The Janus-like faces of these two diseases suggest that the striatum and the substantia nigra normally operate in balance. If the balance of power becomes one-sided, then there is either too much movement (Huntington's chorea) or too little (Parkinson's disease). This interpretation is born out by the effects of medication: drugs used to treat Parkinson's disease make Huntington's chorea worse, and vice versa.

But the balance of power is not just between these two regions. For example, damage to yet another part of the basal ganglia causes writhing movements of the hands, rather like an exaggerated form of Lady Macbeth's hand washing. Recently, this complex balance of power between the different parts of the basal ganglia has been exploited in a new treatment for Parkinson's disease that circumvents the need for drugs but puts the patient under the knife. If certain structures connected to the substantia nigra are destroyed during surgery, then, amazingly, there is an impressive recovery of movement in many Parkinsonian patients. So it cannot be the case that the substantia nigra – which is almost extinct in the extreme cases of Parkinson's disease that submit to surgery – is the centre for movement. If movement has reappeared, surely the substantia nigra cannot have a monopoly. Somehow, the interrelations between the relevant brain regions mean that the whole is more than the sum of the parts.

Even the basal ganglia collectively do not constitute a centre for movement, for other parts of the brain play a key role. One such part is the cerebellum ('little brain'), a cauliflower-shaped structure at the back of the brain. Whereas the basal ganglia appear to be involved in primitive types of spontaneous, internally triggered movement – standing up, sitting down, starting to walk – the cerebellum is involved in movement coordinated with input from the senses. Patients in whom the cerebellum is damaged appear clumsy. They cannot coordinate a complex sequence of movements that rely on constant, ongoing inputs from their senses, such as playing the piano or walking along a straight line. The types of movement that depend on the cerebellum are closely linked to the senses; those

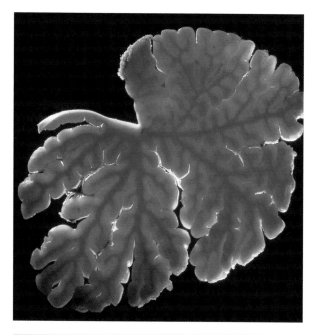

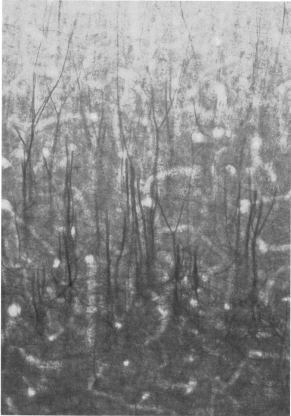

controlled by the basal ganglia arise internally and not as an interplay with what our eyes and ears are telling us.

We can see the different contributions of the cerebellum and the basal ganglia in patients with damage to one or other region. Patients with cerebellar damage can still generate internally driven movements – although they appear clumsy and uncoordinated, they can stand up, walk, and move around. By contrast, a patient with faults in the basal ganglia cannot translate the intention to stand up into a movement. Yet, if a car is rapidly approaching, they can gauge the distance and leap out of the way, as their undamaged cerebellum is pressed into service. In Parkinson's disease, the cerebellum remains unaffected, so one way of treating the condition is to try and transform movements that depend on the substantia nigra into movements that depend on the cerebellum. For instance, if cardboard footprints are placed on the floor, then a Parkinsonian patient can walk more easily by trying to follow the footprints. By using feedback from the senses to control where they walk, the patient bypasses their deficient substantia nigra and makes use of the healthy cerebellum.

Movements such as standing up and walking – and even complex skills involving the senses, such as playing tennis or driving a car – seem to happen unconsciously. We do not consciously have to select the sets of muscles that must contract and relax in an orchestrated sequence to produce the desired movement; instead, the cerebellum and basal ganglia liberate us from the need for

Recent brain-scanning techniques give us powerful tools to watch the brain at work, without the need for surgery. Two of the new techniques exploit the fact that our brains are very greedy for energy. The human brain is a small fraction of total body weight, but it demands fuel – in the form of oxygen and glucose – at a far greater rate than any other organ. Oxygen and glucose are carried to the most hard-working parts of the brain by myriads of blood vessels. The harder-working the region, the greater its consumption, and the greater the blood flow to that site.

PET (positron emission tomography) scanning monitors how much oxygen or glucose different parts of the brain are using. The subject lies down and a doctor gives them an injection of oxygen or glucose that has been tagged with radioactivity. A scanner, which consists of a circle of sensors that detect radioactivity, is then placed around the subject's head. The sensors pick up which parts of the brain have the most radioactive oxygen or glucose, and this is displayed as a colour image on a computer screen. The most active areas of the brain literally light up in colour.

The technique of fMRI (functional magnetic resonance imaging) also monitors which parts of the brain are using energy most quickly. In fMRI, the scanner measures oxygen levels by picking up faint radio signals emitted by haemoglobin, the oxygen-carrying pigment that gives blood its colour. An advantage over PET scanning is that no injection is required. The entire body is slid into a giant, tube-shaped magnet.

Both PET and fMRI are indirect – they measure the increase in blood flow that occurs when a part of the brain is particularly active. By contrast, MEG (magnetoencephalography) monitors the activity of the brain cells themselves. Brain cells generate electrical blips that act as signals to neighbouring cells and the MEG scanner detects the magnetic fields produced by these blips. During the scan, the subject sits in a chair in a room screened from all electrical interference. A helmet is lowered into position so that the head is surrounded by an array of SQUIDs, or 'superconducting quantum interference devices'. These are incredibly sensitive. The magnetic fields they pick up are about a billionth of the strength of Earth's magnetic field, and a hundred times weaker than the magnetic field surrounding a household wire.

Before the invention of MEG scanners, electrical activity in the cortex could be monitored by electrodes placed on the scalp. This technique, in use since the 1920s, produces a printout of brain wave patterns called an EEG (electroencephalogram). Although valuable for revealing the changes in brain waves that occur in sleep, dreaming or in dysfunctions such as epilepsy and coma, electroencephalography is a crude tool. It cannot tell exactly what different parts of the cortex are doing at precise moments. MEG is a great advance in this respect as it can pinpoint tiny areas of activity very precisely, and over timescales much shorter than is possible in PET or fMRI scanning.

The problem with MEG, as with its more established forerunner, is that the best recordings are obtained from the cortex; signals begin to fade out on looking deeper into the brain. While PET and fMRI give the whole picture, they do so on a much more sluggish time scale – seconds, as opposed to the milliseconds in which the brain actually works. A supporter of MEG once described PET and fMRI as studies of 'the bored brain'.

Despite its limitations, MEG certainly seems able to show us the brain working at full tilt. On a screen showing a MEG recording of a grey-coloured brain, tiny splotches of colour – active areas – appear and vanish in a trice. Splotches come and go all over the cortex, like some neuronal firework display. And, I am told, it is never the same twice. If you tried to average the responses over a longer timeframe, you'd be back to the 'bored brain' again, the routine ticking over rather than the one-off crackle and pop of a single, unique moment.

So the dilemma is that PET and fMRI are too slow to capture the fine-tuning of brain activity, and MEG cannot see deep enough to reveal the interplay of different brain regions. In the future, we need to get round these limitations and combine the best of all these techniques.

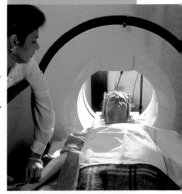

▲ A radiographer injects a patient about to undergo a PET scan.

◄▲ A section through the cerebellum of a 1-year-old infant.
◄◄ Light micrograph of a longitudinal section shows neurons in a line in the cortex, with processes extending upwards towards the outer edge of the brain.

conscious control. In the case of the cerebellum, the subtle interplay of senses and movement has to be learnt – as anyone who has struggled with learning to drive a gearshift car or play the piano well knows – but, eventually, we are left to think about other things as our body effortlessly does what we want.

However, there is another dimension to movement beyond the realm of the cerebellum or the basal ganglia. And that concerns the conscious *will* to move. What part of the brain actually decides when, where, and how to move? The answer lies buried somewhere in the cortex, the outer, folded part of the brain that is so unusually large in humans. Here, we find yet more regions that play vital roles in the control of movement.

By placing electrodes on the scalp to detect the faint electrical signals, or brain waves, made by different parts of the brain, scientists can obtain a glimpse of what goes on in the cortex when a person makes a conscious decision to move. A good second or two before the movement, there is a change in the pattern of brain waves produced by the upper back part of the cortex (the parietal cortex) – almost as if the brain were saying 'Get ready'. Then, a second brain wave starts off at the front of the cortex – 'Steady', says the brain. Finally, as the movement is unfolding, a third brain wave buzzes from just behind the front of the cortex – 'Go!', orders the brain.

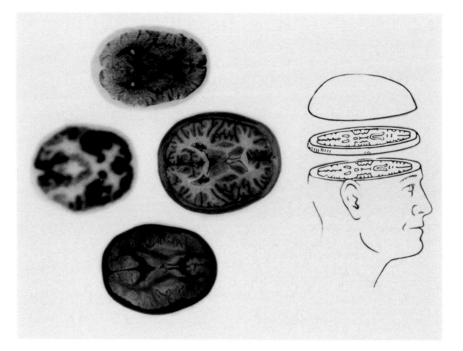

▶ The four techniques for imaging a slice of the brain. Top, X-ray computed tomography (CT); centre left, positron emission tomography (PET); centre right, standard photography; bottom, nuclear magnetic resonance (NMR).

The most intriguing question, of course, is where does the actual *will* to move come in? And exactly who has decided to move – you or your cortex? We cannot say that the parietal cortex, which seems to set in train the intention to move, is the centre for free will because it, in turn, has been receiving signals from a host of other brain regions. In fact, the decision to move cannot be pinned down to just one bit of brain – it is the final product of memories, feelings and information coming in from the senses. Just where and how 'we', or our brains, take that first mental step towards a movement – indeed the very issue of who is in control – are questions that we shall be exploring right up to the final chapter.

Although neurologists have long known that there is no such thing as a single centre committed to each specific mental function, the extraordinary degree to which many brain regions seem to work together became apparent only with the advent in recent years of powerful imaging techniques, which have revolutionized brain research. These imaging techniques – there are several – have in common the ability to reveal the human brain at work, painlessly, while a person is conscious.

Imaging was an immediately appealing way to study the brain because it meant that, finally, neuroscientists could study the living human brain directly and so reduce the amount of controversial research performed on animals. Moreover, the scientists could now carry out research that was impossible with animals, such as the study of language. And the study of tasks involving complex instructions and subtle responses, which would at best have required extensive training in animals, became much easier.

One of the most fundamental findings to come from imaging studies has been that no single area of the brain is activated exclusively and solely during a specific mental task. On the contrary, whole constellations of brain regions become active at different times and in subtly different configurations, depending on what the person being studied is doing. Indeed, far more brain regions turn out to be involved than had previously been thought possible.

But these new techniques, revolutionary, intriguing and aesthetic as they may be, are still not the answer to a neuroscientist's prayer. Brain imaging is a pure case of 'what you see is what you get'. Just knowing that a certain part of the brain is active to a certain extent during a specific task does not actually inspire any new theories as to what is going on, nor how or why. We simply know that a certain brain region is active – the underlying scheme can never be revealed by brain imaging techniques alone.

To illustrate my caution about how much brain imaging can really tell us, let's return to the neurodegenerative diseases. Long before brain imaging, the two crucial areas of the brain involved, the substantia nigra and basal forebrain, were already familiar. Even the deficient chemicals, dopamine and acetylcholine, were understood well enough to lead to the development of drugs that boost the level of these chemicals. But we still do not know why a lack of dopamine makes it so hard for Parkinson's patients to translate their thoughts into movement. And the action of acetylcholine on brain cells is equally difficult to relate to the tragic dismantling of personality that takes place in Alzheimer's disease.

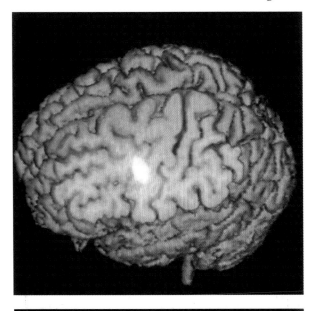

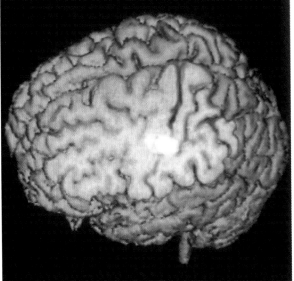

Just as the idea of the brain being divided into centres for specific functions turns out to be too simplistic, so too does the idea of neurotransmitters being chemicals for specific mental tasks. Let's return to the example of dopamine, the absence of which in Parkinson's disease causes loss of the ability to move. It might seem that dopamine might be the chemical for movement – after all, raising its level with the drug L-DOPA gives back patients the ability to move. But dopamine is not produced only in the substantia nigra. Just next to this part of the brain is another dopamine-producing region, the ventral tegmental area, and a disturbance in dopamine level here leads not to Parkinson's disease but to schizophrenia.

Contrary to popular belief, schizophrenia is not a condition of split personality. Rather, it is a split with what most of us would construe as reality. Actual diagnostic criteria for schizophrenia vary enormously, and some regard the term as an umbrella label for a variety of related but distinct conditions. Nonetheless, there are core symptoms, such as hearing voices talking about oneself, and the belief that somehow one's thoughts are independent of one's brain – for example, a schizophrenic may believe that a thought has been inserted or downloaded into their head by an alien agency.

Schizophrenia is different from the neurodegenerative diseases, not only in its florid and complex symptoms but also in its apparent cause. Unlike Parkinson's disease and Alzheimer's, schizophrenia does not seem to be caused by physical damage to a key brain region. Symptoms can reverse, or reoccur, neither of which fits with a theory of enduring damage to a part of the brain. The bizarre way of thinking and talking, which reflects a view of the world very different from that of most people, is perhaps not so much anatomical as chemical.

In schizophrenia, the problem seems to involve dopamine in some way. But whereas Parkinson's disease is characterized by too little dopamine, in schizophrenia there seems to be too much. Excessive dopamine was first found to play a role in schizophrenia when it was discovered that one apparently successful treatment is the use of drugs that block the action of dopamine. This strategy is the very opposite to that used to treat Parkinson's disease, and this has some interesting consequences. If a schizophrenic patient takes drugs to counter dopamine, the blockade is effective not only in the region where dopamine is produced in excess, but also – because the drug is taken orally as a tablet – in the adjacent brain region, the substantia nigra, which until now had been functioning quite normally. Once the drug gets to work here, it causes the same impairment of movement that occurs in Parkinson's disease. Similarly, when a Parkinsonian patient is treated with L-DOPA, the drug boosts dopamine levels in the substantia nigra but also acts in the neighbouring area, where it is not needed. Too much dopamine is produced, with the result that a side effect of L-DOPA may well be the type of hallucinations and delusions that characterize schizophrenia.

Another way people can give themselves schizophrenia-like symptoms, such as paranoia, is to take a large dose of the stimulant drug amphetamine, which works by increasing the availability of dopamine in the brain. The connection between dopamine and schizophrenic symptoms originally led brain researchers to think that schizophrenia is caused by an excess of dopamine. Needless to say, the picture is far from being that simple. We still have a lot to learn about the subtle interplay of dopamine with other neurotransmitters and other brain regions, and how this interplay might somehow lead to the disturbed thought processes seen in schizophrenia. And just as any one dysfunction may involve more than one neurotransmitter, so any one neurotransmitter, like dopamine, may be involved in more than one dysfunction.

◀◀ MEG scans showing brain–hand control. Above: part of the motor cortex is activated a thousandth of a second before a patient starts moving their right index finger. Below: within 40 milliseconds this activation of the fingers is processed in the adjacent area just behind in the somatosensory cortex (concerned with the processing of touch, pain and temperature). Only with MEG scanning can this 'real time' degree of precision, in this case less than a tenth of a second, be visualized.

So dopamine cannot be the chemical 'for' movement. How could a tiny molecule on its own have a function, or dysfunction, locked away inside it? As we have seen, dopamine has very different effects according to where it acts. Similarly, in Alzheimer's disease, acetylcholine can hardly be regarded as the chemical for memory or personality. Other parts of the brain use the same neurotransmitter and remain unaffected by the disease. Moreover, some of the parts of the brain that degenerate in Alzheimer's do not use acetylcholine at all, and instead use other types of neurotransmitter.

So, just as we saw that certain brain regions play an essential part – but only a part – in various functions and dysfunctions, so also with neurotransmitters. In order to understand Parkinson's and Alzheimer's diseases, we need to look beyond the mere neurotransmitter. One possible clue might come from the well-established fact that the two pathologies often coexist in the same patient.

In the past, most brain scientists and clinicians regarded Alzheimer's and Parkinson's diseases as different disorders, each producing its own characteristic lesions in the brain that could be identified under a microscope. We are now realizing that the distinction may not be so clear-cut. The lesions produced by these two diseases share certain chemical components in common, and they sometimes appear in the part of the brain normally associated with the other disorder. The identification of either Parkinson's disease or Alzheimer's disease with a certain type of microscopic marker is thus becoming blurred.

As a result, some scientists are now starting to focus on what Alzheimer's and Parkinson's diseases have in common. Perhaps there is a common factor that triggers the death of brain cells in both conditions. After all, progressive degeneration is not normal for brain cells. In most cases of localized brain damage, such as stroke, the brain is able to recover to some extent. Strokes are caused by sudden disruption in the blood supply to a part of the brain, often due to a blocked or ruptured blood vessel; brain cells in the vicinity are starved of oxygen and die, causing various impairments, such as a paralysed arm or sudden loss of speech. Yet, in many cases, there is a certain degree of recovery from a stroke, sometimes even a complete compensation by adjacent parts of the brain. Only when the damage through stroke or some other malfunction is in the substantia nigra or the basal forebrain, or in other key places deep within the brain, does Parkinson's or Alzheimer's disease result. There must, therefore, be something special about these groups of neurons that sets them on a continued path of self-destruction.

The critical quest now is to discover what that 'something special' might be. If it turned out to be a common factor shared by the cells of the substantia nigra and the basal forebrain, its discovery could lead to the development of much more effective drugs. Parkinson's and Alzheimer's diseases are currently treated with different drugs, and in both cases the aim of therapy is to relieve symptoms by temporarily replenishing the deficient neurotransmitter. But no therapy can yet tackle the fundamental problem – perhaps the common factor – that causes the cells to die. In order to arrest the degeneration, we need to know *why* the cells die in the first place; only then will we be able to develop an interceptive strategy to stop cell death.

We have seen in this chapter that there are two ways of trying to understand how the brain works. We can look at the big picture, investigating the role of large-scale brain regions and how these interact with each other. Or we can look at the nuts and bolts of the brain, by studying how chemicals such as drugs and neurotransmitters work. Before we can reconcile these two levels, we first need to find out what goes on inside any one brain region, and to do that, we need to know what the brain is actually made of.

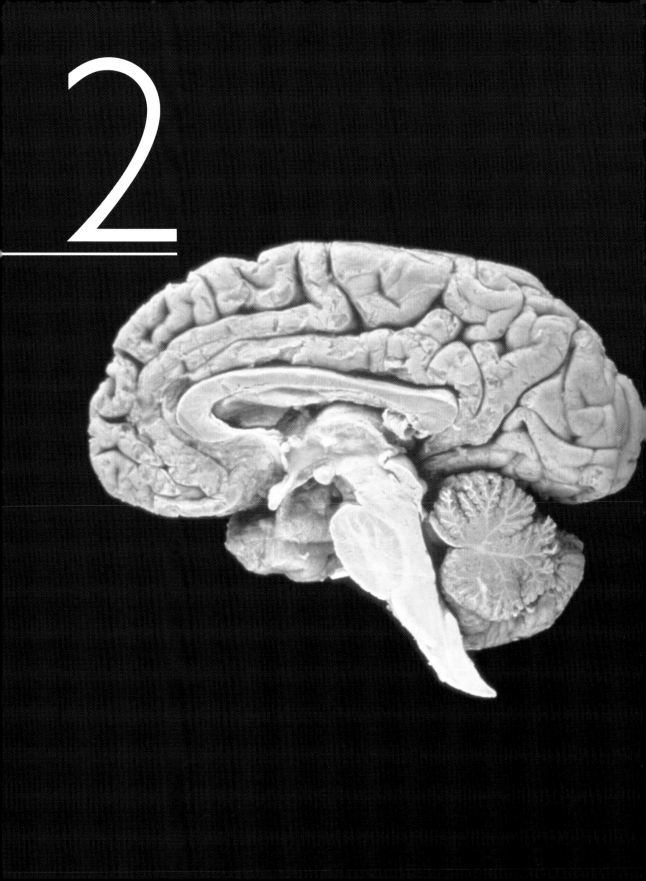

CEREBRAL NUTS AND BOLTS

I remember the scene as if it were yesterday – the dissection room in the Department of Anatomy at Oxford University, with rows of benches, and us students perched along in pairs in our stiff white coats, flexing our hands in their unfamiliar carapace of surgical gloves. The white plastic pots being dumped unceremoniously in front of each pair; the clipped, impersonal instructions from the front; the nervous glance at the textbook at my elbow. Then, finally, prising off the plastic lid and reeling at the evil fumes of the chemical fixative. For inside the pot was a human brain. My partner and I started the stages of identification and dissection required. But it didn't take long before I had to pause. I was holding the very essence of a person. Although lifeless, this pickled object still, in some sense, held a person's memories, fears and dreams. Would a bit scraped under my fingernail – if I had not been wearing gloves – contain a particular memory or dream? And, if so, how could such unpromising creamy stuff do all that? What *was* this creamy stuff anyway?

Around 400 BC, the Greek philosopher Democritus argued that there had to be a physical basis for the mind. He took the view that everything in the universe was composed of minute, invisible particles, which he called atoms (from the Greek for indivisible). According to Democritus, atoms not only made up matter – which is consistent with our modern view – but also were the stuff of less tangible entities, such as mind. In a sense, even here Democritus was correct: if you took the brain apart to reveal its smallest working components, its basic building blocks, you would end up with cells. You could even pick these cells apart into molecules, and then pick those molecules apart into atoms. For now, though, we shall stay at the intermediate stage of dismantling the brain and concentrate on its cells. It is the existence of these brain cells, much larger than atoms but still invisible to the naked eye, that underpins the workings of the brain.

◀◀ Slicing through the human brain: the front of the brain is on the left, with the folded cortex at the top and the elongated brain-stem at the lower centre. The cerebellum, or 'little brain', is at the back, lower right.

Nearly everyone has heard of the term neuron. The neuron is the main type of brain cell, and it plays a centre-stage role in what goes on in the brain. The pivotal job of the neuron, as we shall see in more detail shortly, is to send very rapid messages to other neurons. These messages are just electrical pulses and sprays of neurotransmitter, yet they somehow build up into the complex and sophisticated mental feats that fascinate not just the majority of brain scientists, but anyone contemplating the workings of their own brain.

CINDERELLA CELLS

Our brains contain a mind-boggling 100 billion or so neurons, but there are many more cells besides these. The other cells are glial cells, which are vital to the healthy working of the brain. These take their name from the Greek word for glue because many appear to stick to neurons or blood vessels. Glial cells are sometimes thought of as the poor relations of neurons. In the best tradition of poor relations, these Cinderella cells are occupied primarily with housekeeping. And like workers everywhere, they outnumber those they serve, in this case the neurons – there are a staggering 1 trillion glial cells in the human brain – ten times the number of neurons.

Some glial cells play an important role in repairing damage. These cells can actually move about in the brain like vultures, homing in on damaged tissue and scavenging debris from injured cells. Another type of glial cell lines the inner surface of several interconnected cavities, called ventricles, that exist in the centre of the brain. These cavities contain a watery liquid known as the cerebrospinal fluid (CSF), which bathes the brain and the spinal cord and protects them from injury. The CSF was once called 'the urine of the brain'; like urine, its chemical composition changes in people with certain diseases, such as brain tumour, so the CSF is very useful in diagnosis. To obtain a sample for analysis, doctors insert a hollow needle into the spine, a procedure called lumbar puncture.

The most common glial cells are the astrocytes, which are named for their star-like shape. Astrocytes cluster all around the neurons like dutiful attendants. In the past, brain scientists tended to dismiss astrocytes as having a rather dull, supporting function, acting merely as a kind of biological scaffolding to prevent the neurons from slithering around. We now know that astrocytes are much more actively engaged in the dynamic working of the brain. In a healthy adult, they maintain an ideal chemical 'microenvironment' around each neuron, acting as a sponge to mop up excessive or potentially toxic chemicals. When a neuron is damaged, its astrocytes redouble their efforts, increasing in size and number and releasing the substances needed for it to regrow and repair itself.

The last main type of glial cell is the oligodendrocyte, named after the Greek words for few (*oligos*) and tree (*dendros*). They are so-called because they have relatively few branch-like tentacles emanating from them, unlike astrocytes, which have many. Nonetheless, the few tentacles are put to invaluable use: they snake out to make contact with a handful of neurons and wrap around them several times to form a protective sheath. Like the plastic coating around an electrical cable, the sheath insulates the parts of the neuron that carry electrical signals. Degeneration of these sheaths is what causes the gradual paralyis and other symptoms of the disease multiple sclerosis.

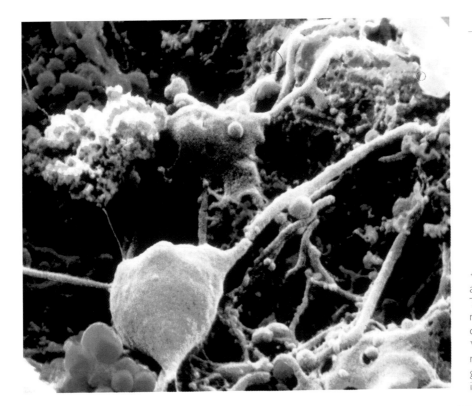

◀ Brain cells as seen in a scanning microscope. This micrograph shows neurons in grey and glial cells in orange.
▼ The Golgi-Cox staining method. A light micrograph showing neurons in the cortex.

It was the Czech anatomist Jan Purkinje (1787–1869) who was first to observe neurons under a microscope, but it was not until 1872, when the Italian physician Camillo Golgi (1843–1926) developed a stain that made neurons stand out, that the beauty and diversity of their delicate shapes was finally revealed. The Golgi stain makes neurons appear black against an amber background; for reasons that are still obscure, it stains only a random 10% or so of the neurons. This is a great advantage – if all the neurons were stained, the brain would appear as a featureless black mass under the microscope.

So imagine you are looking down a microscope at a slither of brain tissue stained by the Golgi method. The stained neurons show up as squat circles or uneven triangles with long, thin branches emanating from them. Using an ordinary microscope with a magnifying power of

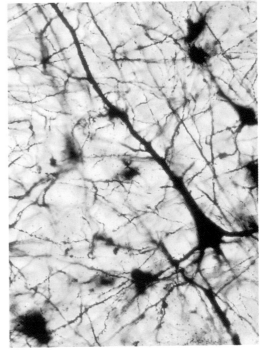

only 1500, these branches appear to connect the neurons together. Camillo Golgi formed the view that the brain was a vast, continuous network of neurons, like a fishing net. This 'reticular theory' came under fire from the Spanish anatomist Santiago Ramón y Cajal (1852–1934). Golgi's scheme of things, according to Cajal, was completely misguided. Cajal proposed that neurons were not continuous at all, but were separate modules that happened to be very close to each other. His view gave rise to the 'neuron theory', in which each neuron is seen as an autonomous unit, distinct from other neurons. Despite their bitter disagreement, Cajal and Golgi were jointly awarded the 1906 Nobel Prize for Medicine and Physiology for establishing the neuron as the basic building block of the brain.

We now know that it was Cajal who was, for the most part, correct. The idea that neurons are discrete entities was proved once and for all in the 1960s, when the advent of the electron microscope made it possible to examine brain cells at magnifications of up to 80,000. It was immediately obvious that neurons are indeed separate entities. Although they come very close to each other, there is still a very small gap at the apparent point of contact. This gap between cells, called a synapse, is very narrow – about a 200,000th of a millimetre across – so it is hardly surprising that Golgi and his followers had thought neighbouring cells were continuous.

But what of the shape of the neurons themselves? Think of them as tiny black boxes, with inputs and outputs and something happening in between. The inputs are thin, wiry branches called dendrites, of which there may be up to 100,000 on a single neuron. The dendrites receive incoming signals from neighbouring cells and relay these signals to central control – the cell body (one of the squat circles and uneven triangles that Golgi saw under the microscope). Once a signal arrives at the cell body, it may be suppressed or amplified by other signals zooming into the neuron. Eventually, as a result of all the incoming signals converging, a new signal is triggered in the cell body and starts its journey out of the cell via a special exit route. This exit route is another thin, wiry branch, this time called an axon. Unlike dendrites, which resemble true branches in that they are tapered, axons are of a constant diameter, and each neuron has only one. Axons vary enormously in length – the shortest run only as far as the cell next door; the longest stretch from the brain down the spinal cord for up to a metre, making them up to 20,000 times longer than the cell body. At its very end, the axon divides into a number of small branches, each with a swollen tip shaped like a mushroom. The mushroom-shaped tips pass on the electrical signal, via synapses, to connecting cells.

So much for the design of a neuron. If we want to know how the brain works, the next question to ask is: what actually *is* the electrical signal?

All matter is made up of atoms. An atom consists of a positively charged nucleus surrounded by a cloud of negatively charged particles called electrons. In normal circumstances, these positive and negative charges balance, and the atom is electrically neutral. However, in a metal, the electrons are free to move from one atom to the next. This flow of charge is what we recognize as electric current. We do not have metallic wires in our heads, so a different mechanism must be responsible for the electrical signals that buzz through neurons. It turns out that neurons use particles called ions to transport charge. An ion is an atom that has become positively or negatively charged by either losing or gaining electrons.

Think of a canal lock, which holds back water when the gates are shut. The water cannot flow, yet the potential is there. Exactly the same principle is put to use by neurons. If the movement of an ion is prevented in some way, it will have a potential for producing an electric current once the lock gates are opened. If there is an accumulation of ions of one sign in a certain region and a deficit elsewhere, or perhaps an accumulation of ions of opposite charge, there will be a potential difference between the two sites. The potential difference is measured in volts, such that the higher the potential difference the greater the current when finally it is allowed to flow. Even though it might not be sending a signal, any living neuron is constantly generating a potential difference – a voltage – as a result of the distribution of ions across its membrane. This potential difference is called the resting potential of the cell and inside it is typically some 80 thousandths of a volt negative with respect to the outside.

The electrical signal which any one neuron fires off is known as an action potential. This is what happens.

When the cell is at rest, sodium ions are at high concentration outside the cell and potassium ions are at high concentration inside the cell. However, for an action potential to be generated, tiny channels open up in the membrane wall and allow sodium ions to flood into the neuron; because sodium carries a positive charge, the inside of the neuron will now be more positive, and so the potential difference across the membrane will be less.

As the potential difference becomes more and more reduced, a second channel, this time one for potassium ions, starts to open up. Potassium is normally inside the neuron, so this channel allows the imprisoned ions to

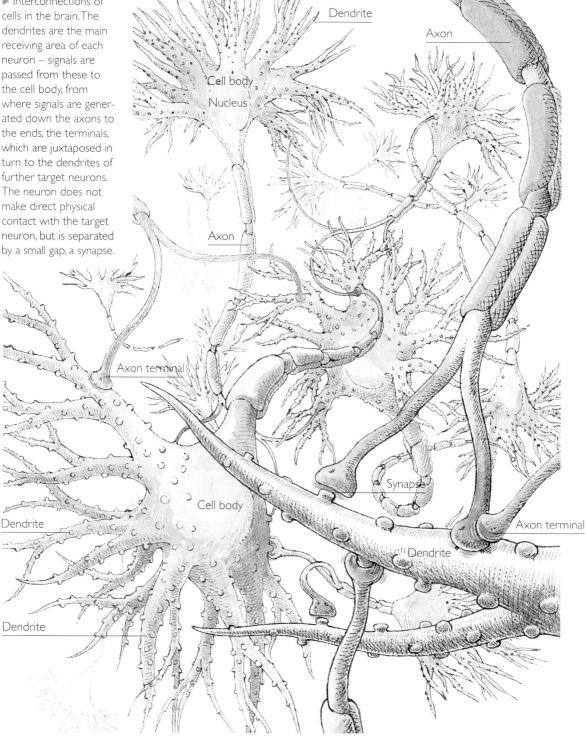

▶ Interconnections of cells in the brain. The dendrites are the main receiving area of each neuron – signals are passed from these to the cell body, from where signals are generated down the axons to the ends, the terminals, which are juxtaposed in turn to the dendrites of further target neurons. The neuron does not make direct physical contact with the target neuron, but is separated by a small gap, a synapse.

Dendrite

Axon

Cell body

Nucleus

Axon

Axon terminal

Synapse

Dendrite

Cell body

Dendrite

Axon terminal

Dendrite

Dendrite

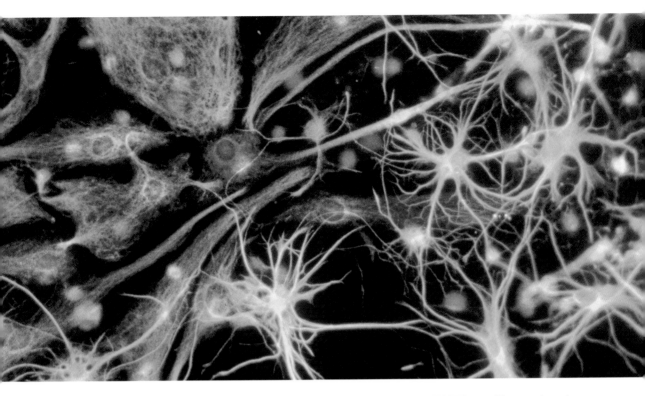

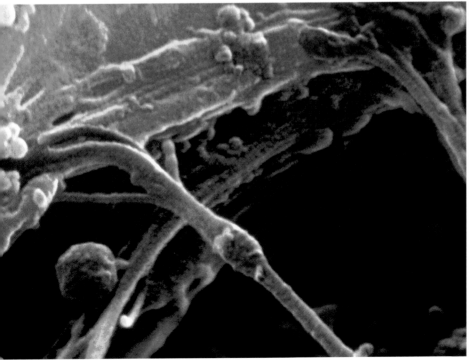

▲ The star-shaped astrocytes, shown here in light green, which supply nerve cells with nutrients.

◄ Scanning micrograph showing synaptic junctions, or synapses, where neurons and connections meet. The nerve fibres (purple), run diagonally. At the top is the surface of the target cell (yellow). The fibres branch and swell before forming the junction and it is at these terminal swellings (top right) that the transmitter chemicals are released on to the cell.

flood out through the membrane in the opposite direction to the sodium ions. For the brief period that the potassium channels remain open, this sudden exit of positively charged ions results in the interior of the cell again becoming negative, like people pushing through an emergency fire door: as a result the potential difference reverts back to being more or less what it was, although in fact there is a slight overshoot. However, molecular pumps embedded in the wall of the neuron quickly return the concentration of the ions to normal, and the cell recovers its erstwhile resting potential. This sequence of events – a surge of positive charge inside, an overshooting return of positive charge outside, and then a return to resting potential – is reflected in the characteristic shape of the action potential, which we can monitor on an oscilloscope.

So the electrical signal, the action potential, consists of a highly transient swapping over of ions inside and outside the cell, and therefore a change in the potential difference, the voltage. It all takes typically 1–2 milliseconds. The more action potentials a cell generates, the harder it can be regarded as working to communicate with other neurons – the more it is 'excited'.

The action potential is now ready to be passed on to the next neuron. From the main part of the neuron, the cell body, it can buzz along the output path, the axon, a little like a domestic electric current along a cable. But unlike the robust copper conductors in the outside world, in the world inside our heads the axons are far from perfect conductors. Although they are insulated with myelin (from one of the types of glial cell mentioned a little earlier), a big risk is that current will still eventually leak, and the action potential diminish to nothing before it reaches its destination. Nature needed a way to prevent this scenario and enable the action potential to arrive at the end of the leaky axon without decrement. The brilliant solution has been to have custom-made breaks in the myelin insulation; at these points, because there is no myelin, the wall of the axon acts like the membrane of the original cell body, and the action potential can be generated afresh, so it effectively jumps along down the axon. This all happens at a surprising speed – up to 400 kph (250 mph)! However, even though the problem has been solved of stopping the action potential leak away, another difficulty looms.

The action potential arrives, at full strength, at the end of the line, the axon terminal. But the axon terminal is the most remote part of the neuron. Once we arrive here it is like having travelled a road that leads no further but is terminated by a river. After the axon terminal, there is the fluid-filled gap of the synapse – narrow maybe, but a barrier nonetheless to an electrical signal. How can the message get across, then, to the next neuron?

An axon terminal contains tiny packages of chemicals that act as chemical messengers: neurotransmitters. Some of these packages, or vesicles, are right up close to the edge of the synapse. The arrival of an action potential triggers the vesicles to fuse with the wall of the nerve terminal itself, thereby emptying out their neurotransmitter contents into the narrow gap between the neurons. About 1–10 vesicles are released at a synapse, with each vesicle containing up to about 10,000 molecules of neurotransmitter.

The conversion of the original electrical signal to a chemical one in this way is a little like swapping a car for a boat. The chemical neurotransmitter can diffuse easily across to the input zone, the dendrites, of the next neuron. There it joins in a kind of molecular handshake with a specialized protein, known as a receptor, on the outer wall of the target neuron. This molecular handshaking leads to the opening of an ion channel, such as the one for sodium ions that we saw is the first phase in the generation of action potential, and a new electrical signal is triggered in the second neuron.

So everything we think and feel can ultimately be boiled down to this alternating sequence of electrical and chemical events. The electrical signal arriving along the axon is converted into a chemical signal that carries it across the physical barrier, the synapse, between the neurons. Then, once the signal reaches the receiving neuron, the chemical signal is subsequently translated back into an electrical one that can travel into the cell body of the new cell. Brain researchers refer to this entire process as synaptic transmission.

When I first learned about synaptic transmission, the lecturers simplified the story by drawing neurons in a simple daisy chain; it was easy to imagine the signal passing along from one cell to the next. In reality, each synapse has a minuscule influence on the target cell, because that cell is receiving tens of thousands of inputs all over its many dendrites. So there is an enormous convergence of inputs into any one neuron, and these end up being averaged at the cell body into a single output – a final action potential that will itself become one of tens of thousands of inputs converging on the next neuron.

The apparent similarity between neurons and electrical circuits led scientists in the 1960s to liken the brain to a computer. The idea was that neurotransmitters could either *excite* neurons (encourage action potentials) or *inhibit* them (prevent action potentials) – two alternative states that seemed analogous to the noughts and ones of binary computer code. If the brain was truly digital, then we might expect to find just two neurotransmitters in brain cells: a positive excitatory one, and a negative inhibitory one. We now know that this scenario is far too simplistic. Unlike a computer, the brain not only makes use of on/off switches, but also uses a surprisingly wide range of neurotransmitters to operate those switches. In other words, it exploits qualitative factors as well as quantitative factors, and this adds a powerful extra dimension to its processing power. As the years have unfolded, more and more neurotransmitters have been discovered, each with a very different profile, and it has slowly dawned on scientists just how incredibly complicated the human brain really is.

The rest of this chapter looks in more detail at those neurotransmitters and the nuts and bolts of how they work.

The first discovery of a neurotransmitter took place shortly before World War I, when the chemist and pharmacologist Henry Dale found that a substance released from nerves in the spinal cord made muscles contract. The substance – acetylcholine – was the very neurotransmitter that we now know is deficient in Alzheimer's disease. As well as acting between neurons in the brain, it serves as the chemical mediator between muscle cells and nerves throughout the body.

For the most part, brain cells that use acetylcholine are located deep in the brain, from where axons fan out to connect with the cortex and a region below it, the hippocampus. As well as being involved in Alzheimer's disease, this 'fountainhead' of acetylcholine plays a key role in sleep, wakefulness and arousal (how excited we feel). We know this because if a person takes the drug hemicholinium, which stops production of acetylcholine, they experience a sharp reduction in the amount of dreaming sleep.

We can tell when a person is dreaming by placing electrodes on their scalp and monitoring their brain waves on an electroencephalogram (EEG) – dreamless sleep produces slow, regular brain waves, whereas dreaming results in much faster, irregular waves. Dreaming also causes rapid eye movement (REM), and so is also known as REM sleep. The slow wave of dreamless sleep is triggered not only by the drug hemicholinium, but also by the drug atropine, which acts as an imposter, binding to the receptors intended for acetylcholine and so blocking the neurotransmitter's action.

Conversely, drugs that boost the level of acetylcholine trigger the rapid brain waves characteristic of dreams. These drugs work by interfering

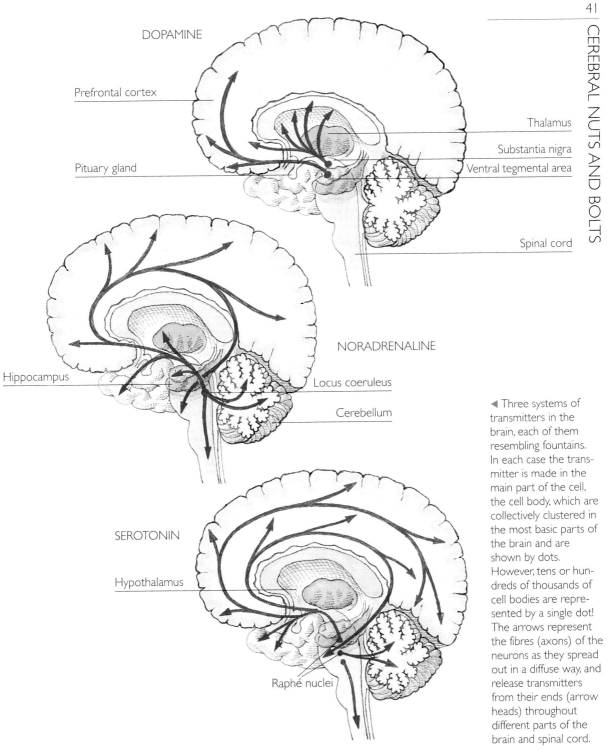

DOPAMINE

Prefrontal cortex

Pituary gland

Thalamus

Substantia nigra

Ventral tegmental area

Spinal cord

NORADRENALINE

Hippocampus

Locus coeruleus

Cerebellum

SEROTONIN

Hypothalamus

Raphé nuclei

◄ Three systems of transmitters in the brain, each of them resembling fountains. In each case the transmitter is made in the main part of the cell, the cell body, which are collectively clustered in the most basic parts of the brain and are shown by dots. However, tens or hundreds of thousands of cell bodies are represented by a single dot! The arrows represent the fibres (axons) of the neurons as they spread out in a diffuse way, and release transmitters from their ends (arrow heads) throughout different parts of the brain and spinal cord.

with a chemical that normally removes acetylcholine from the synapse once it has shaken hands for long enough with its receptor. This chemical, which enjoys the jaw-breaking name of acetylcholinesterase, is an enzyme – a type of protein that carries out a very specific job in the body's complex biochemistry. Interestingly, it is also by blocking the action of acetylcholinesterase that nerve gas achieves its lethal effects. This might seem paradoxical – if acetylcholine is the neurotransmitter that makes muscles move, why should an increase in its level result in fatal paralysis of the whole body? The reason is that an excess of acetylcholine in synapses causes the receptors and their ion channels to go into a kind of molecular stall, so that nothing functions. Even the muscles that control the lungs become paralysed, resulting in swift death by asphyxiation.

The active ingredient of nerve gas is a type of chemical called an organophosphate. It now seems that exposure to even very small amounts of organophosphates might have strange and lasting effects on the body, as cases of organosphosphate-poisoning caused by sheep dip indicate. Organophosphates are also blamed for the Gulf War syndrome experienced by soldiers who have been exposed to small amounts of nerve gas, or who have received vaccinations against nerve gas.

During the early 20th century, scientists were too preoccupied with how chemicals affect the heart and other vital organs to concern themselves with the more daunting task of working out how brain chemicals work. Although Henry Dale discovered acetylcholine before World War I, it was not until the 1950s that three more transmitters entered the stage: noradrenaline, dopamine and serotonin. These three neurotransmitters are, in chemical terms, similar, so it is perhaps not surprising that they work in similar ways, even though their effects are very different.

The neurons that release noradrenaline are, like those that release acetylcholine, organized a little like a fountain. The cell bodies clustering together at the fountainhead form a small part of the brainstem, a stalk of tissue that joins the brain to the spinal cord. The fountainhead contains only about 20,000 neurons, but their axons reach far and wide. As well as running down into the spinal cord, they fan out into the cerebellum at the back of the brain and into large areas of the front of the brain, again including the cortex and hippocampus.

The release of noradrenaline in the brain increases arousal, making one feel wide awake and alert. Drugs such as amphetamine and cocaine, which boost the availability of this neurotransmitter – or even injections of the neurotransmitter itself – can therefore induce an alert, hyperactive state. But noradrenaline has other effects too. By means of the axons

Neuroscientists have devised an ingenious means of detecting the split-second release of certain neurotransmitters from single brain cells. The secret lies in the way these transmitters are rendered inactive after they have carried their message across the synapse. Noradrenaline, dopamine and serotonin are all absorbed back into the interior of the neuron and inactivated by a chemical reaction known as oxidation. Oxidation can be complex, but it boils down to the removal of an electron from a substance.

Because an electron is transferred from one substance to another when oxidation takes place, we can use an electrical technique, voltammetry, to detect the tiny electric current that the process generates.

A miniature electrode is inserted into the appropriate part of the brain. When neurotransmitter molecules are oxidized, the electrode picks up the resulting current and measures it. The amount of current generated gives an accurate indication of how much neurotransmitter is present near the electrode.

Voltammetry is a great advance in brain research because it enables us to detect the release of neurotransmitters in an area about the size of a single brain cell. Readings can be taken more than ten times a second, so the timescale is comparable to the speed of synaptic transmission itself. Another advantage is that separate readings can be taken for noradrenaline, serotonin and dopamine simultaneously.

running down the spinal cord, the noradrenaline fountain influences nervous signals from the brain that are involved in pain and sexual behaviour. Destroying one of these noradrenaline pathways in a female rat can have curious effects. The female spends more time actively soliciting males, but she becomes less able to perform the reflex actions involved in copulation, such as taking up a curved back posture. Clearly, noradrenaline plays a role in a wide range of very different types of behaviour.

Noradrenaline is made from the chemical dopamine. As we saw in chapter one, dopamine is a neurotransmitter in its own right, produced by the black 'moustaches' of the substantia nigra. As well as being involved in movement, and playing a part in schizophrenia and Parkinson's disease, dopamine is also found in sense organs – in particular the eye and the circuits of neurons leading from the nose. It also serves as an important neurotransmitter between the brain and the pituitary gland, a pea-sized gland that dangles on a stalk from the underside of the brain and produces important hormones affecting growth, sexual development and other body processes.

A chemical cousin of dopamine and noradrenaline is serotonin, which also originates from a fountainhead of neurons at the head of the spinal cord. The cell bodies of the serotonin fountainhead form small clumps that are arranged in a line along the brain's central axis. Some of these clumps have axons that reach right down into the spinal cord; others connect with the hypothalamus, a small region in the base of the

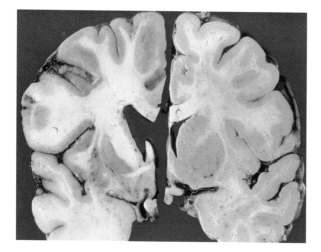

▲ The effects of Huntington's chorea: on the right, a slice from a normal brain; on the left, a section from a Huntington's chorea patient showing severe degeneration of the caudate nucleus and enlarged ventricles, due to the loss of surrounding tissue.

brain concerned with sex, temperature, hunger and thirst. Yet more serotonin-releasing neurons link up with the amygdala, an almond-shaped region deep in the brain that is involved in emotion. And there are also serotonin pathways to the cortex (the site of 'higher' mental functions), and to the striatum and cerebellum, both of which are involved in movement.

In view of the brain's many and varied serotonin pathways, it should come as no surprise that serotonin is linked to all sorts of mental processes. Two are particularly important: pain and sleep. We feel pain more acutely following drug-induced depletion of serotonin; destruction of the serotonin fountainhead has the same effect. Conversely, stimulation of the fountainhead or injection of serotonin into the spinal cord both reduce pain. Damage to the fountainhead also leads to insomnia, whereas infusion of serotonin into animal brains induces dreamless 'slow-wave' sleep. As well as being involved in pain and sleep, serotonin seems to play a role in elevating one's mood – the antidepressant Prozac works by increasing the availability of serotonin, as does the drug ecstasy.

Clearly, serotonin cannot be matched directly with any single function. Rather, it is implicated in a *range* of mental processes, and the same is true of the other neurotransmitters. Similarly, in any mental process – be it sleep, movement, pain, arousal or anything else – more than one neurotransmitter is involved.

In the 1960s scientists discovered a new class of neurotransmitters: the amino acids. One particular amino acid – GABA (gamma-aminobutyric acid) – had already been found in brain extracts, but scientists thought nothing of this because GABA was known to be fairly ubiquitous. Granted, it was a neurochemical, but there was no reason to think it might be a neurotransmitter. The big breakthrough came when it was discovered that GABA could inhibit neurons in lobsters.

GABA is present throughout the human body in tiny amounts, but its concentration rises in the brain and spinal cord to about 1000 times that of noradrenaline, dopamine or serotonin. One estimate is that up to 30% of synapses in the brain use GABA as a neurotransmitter. Unfortunately, its precise role is hard to study because it is difficult to distinguish GABA molecules acting as neurotransmitters from those involved in other processes, such as cell housekeeping. Nevertheless, scientists have made interesting discoveries about GABA. For one thing, it seems to be in those parts of the brain concerned with movement. Might GABA, therefore, be the neurotransmitter for movement?

If the situation were that simple, it would be hard to explain precisely how GABA features in the distressing effects of Huntington's chorea. This disease, characterized by wild, involuntary movements of the limbs, occurs as a result of loss of a specific group of brain cells that use GABA as one of their chief neurotransmitters. The excessive movements are related to a *drop* in GABA levels. Could this mean, then, that GABA is the neurotransmitter for *inhibition* of movement?

It turns out that GABA often acts as an inhibitory neurotransmitter. When it binds to its receptor, the usual outcome is that the inside of the target cell becomes more negatively charged, which makes an action potential harder to generate. But the final effect of this inhibition depends on the precise neuronal context, and GABA is involved in many different contexts in the brain, affecting not just movement but other processes as well. Consequently, GABA plays a role in a number of very different mental disorders.

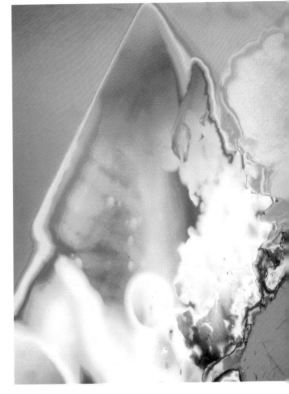

▲ Crystals of the neurotransmitter GABA which has an inhibitory action on brain cells, and which is deficient in the disorder of movement, Huntington's chorea.

Let's take epilepsy and anxiety as examples. Epilepsy is characterized by convulsions, whereas anxiety is more a problem with one's private inner world. It goes without saying that these are very different conditions, but both can be treated by drugs that have essentially the same action – to increase GABA power in the brain. Benzodiazepine drugs, for instance, modify GABA's handshake with its receptor, giving the neurotransmitter a more powerful effect.

GABA inhibits the unwanted symptoms of all these conditions – be they wild, uncontrolled movements, epileptic fits, or wave upon wave of anxiety. So what is the common physical entity in the brain, in each case, that GABA is inhibiting? The answer is that there is no such common factor. GABA works in many different contexts and in many different places, and we need to find out more about these contexts – the complex circuits in which GABA operates – before we can understand how GABA fits in to the bigger picture. What is certain for now, however, is that the molecule itself, independent of context, does not have a particular function locked up inside it.

If noradrenaline, dopamine and serotonin were the neurotransmitters of the 1950s, and amino acids were those of the 1960s, then the talking point for brain scientists in the 1970s were the peptides. Peptides are much larger than amino acids; in fact they are made up of strings of some 10 or more amino acids joined together. Although they were discovered several decades ago, their role in the brain largely remains a mystery. One baffling puzzle is that peptides are often found in the same cells as other neurotransmitters. This makes no sense – the conventional view is that only one neurotransmitter is needed to carry a signal across a synapse. What is the point of a seemingly redundant neurotransmitter?

The Swedish neurochemist Thomas Hokfelt is one of the world's leading authorities on peptides. He has revealed that, although peptides and other neurotransmitters coexist in the same cell, there are important differences in the way they are stored and released. The conventional neurotransmitters are stored on their own in small packets (vesicles) at the end of the axon. In contrast, peptides are stored in bigger packets, which also contain the conventional neurotransmitter and some other chemicals too. When the neuron is ticking over in the normal way, only the small packets of conventional neurotransmitter are dumped into the synapse. But when the neuron becomes more excited, the big packets of peptide start to be released as well. The more excited the neuron becomes, the more the ratio swings from conventional neurotransmitter to peptide. The fascinating issue here is that a neuron can convert a signal relating to *quantity* (the rate of generation of action potentials) into a signal concerning *quality* (the type of chemical).

Another major difference is the site of release. Surprisingly, peptides are released not from the end of an axon into a synapse, but from the sides of an axon. This means that they have a far greater sphere of influence, affecting many surrounding cells. Although we are yet to find out the full implications of this newly discovered chemical mechanism, it is clear that we must abandon once and for all the notion that brain cells work like the parts of a computer.

The 1980s saw the discovery of another type of neurotransmitter that made us question what we understand by the term. Enter the so-called gaseous neurotransmitters, of which the most famous is nitric oxide. To everyone's amazement, this humble little molecule – which in the outside world is involved in the formation of smog and acid rain – has turned out to play an important role in the brain. Chemists are familiar with nitric oxide as a gas (not to be confused with nitrous oxide, or laughing gas), but we should not think of it as bubbling away inside us. Like other chemicals secreted in the body, it dissolves in the fluid environment that surrounds cells. In the brain, nitric oxide affects the immediate neighbours of the neuron secreting it. The discovery of its surprising role as a neurotransmitter won the team of scientists responsible the Nobel Prize for Medicine and Physiology in 1998.

Neuroscientists now know of an enormous range of neurotransmitters, and the list is growing all the time. But even more diversity is possible. Not only are there many different types of transmitter, but there are, in addition, several different sub-types of receptor for any particular neurotransmitter. These sub-types of receptor differ in the amount of neurotransmitter needed to activate them, and also in the ion channels to which they were linked. By this balancing of position, receptor sub-type and neurotransmitter availability, we can start to see how the brain achieves an enormous flexibility in the way neurons communicate. Just as sub-types of receptors give rise to a divergence in the action of any one neurotransmittter, enabling it to affect more than one ion channel, so a convergence of neurotransmitter action can also occur. Imagine two different neurotransmitters each activating its own receptor. If those two receptors are linked to the same ion channel, then the two neurotransmitters would be effectivly working together either to open or to close it and hence contribute to an electrical signal. In this way two neurotransmitters can work in conjunction to influence the excitability of a neuron.

We have to conclude that synaptic transmission is no simple relay of on/off signals from one cell to another. The sequences of electrical and chemical events that enable neurons to communicate are incredibly variable and dynamic. From one moment to the next, the situation is changing, depending on the number, strength, sites and combination of inputs into a neuron that happen to be active. If communication between neurons is indeed the building block of the brain, we can see that even at this basic level there is already the potential for the brain to function on an enormously flexible basis.

But neurons and neurotransmitters are not the end of the story. Every molecule in your brain is there because it has been manufactured in response to a command passed down the generations to you by the all-powerful genes. In the next chapter, we will look at exactly what a gene is, how it works to produce chemicals in our brains, and how those chemicals are influenced by other factors in the environment. To what extent do nature and nurture affect our brains?

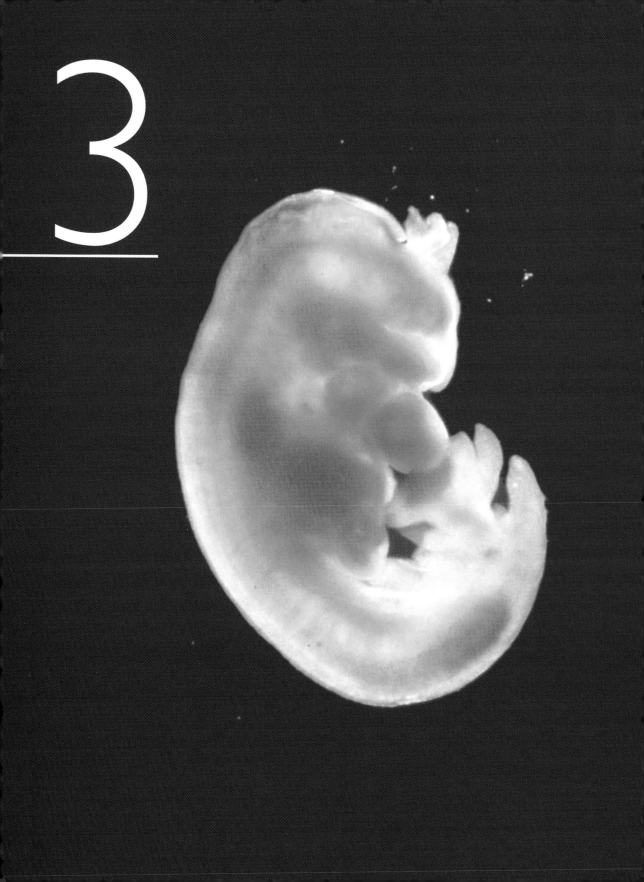

3

NATURE AND NURTURE

The Fiennes family has lived in the same stately home for almost 700 years. As I walked along the portrait gallery, looking at each generation immortalized on canvas, the similarity in the faces was striking. But I could tell very little about what kind of people they were, about how the unique experiences of each individual, and the pressures and prejudices of their particular era, had shaped their minds. The lords and ladies of this ancient family, like all of us, were a product of the genes they were born with and the environment in which they lived. In this chapter we shall investigate how our genes interact with the world around us, influencing the development of our brains and moulding us into distinct individuals.

Our story starts, inevitably, in the womb. Within a day of a human egg cell being fertilized by the father's sperm, it has divided into two cells. Two days later, it has become a ball of 64 cells, and its journey down the fallopian tube comes to an end as it embeds itself into the wall of the uterus. The ball then turns into a hollow sphere, with cells flattened out into two outer layers. By three weeks after conception, these two layers have become three. The middle layer releases substances that act on the outer layer of cells and determine their fate – now they are destined to become neurons and glial cells. The outer layer curls up into a tube; one end starts turning into a brain, the other into a spinal cord. The brain-end swells, developing three bulges that will eventually become the front, middle and back of the brain. At around 40 days after conception, these bulges start to kink away from the developing spinal cord. By 100 days, they have formed a recognizable brain.

During this time, the weight of the brain has increased from zero to some 30 g (1 oz), but now the rate of growth really takes off. By 150 days, the brain weighs almost 80 g (3 oz), and by 190 days it is more than 200 g (7 oz). This will double by the time the baby is born at around 270 days. The growth of the front part of the brain is particularly dramatic. As it expands, it spreads upwards and outwards, covering the rest of the brain.

◄◄ A human embryo after 28 days' development when most of the primitive organ systems are already formed.

At 6 months gestation the human brain looks complete, but there is still much to do. It has effectively been growing inside out, the new neurons forming in the centre and then migrating outwards to take up their positions in the cortex, the brain's outer layer. The new cells are produced at the rate of a quarter of a million a minute, so it is not surprising that the cortex is still able to undergo a radical change in appearance. In the last three months of pregnancy, the cortex loses its smooth appearance and becomes as wrinkly a walnut. The reason for this change is that, as the cortex grows, its surface area expands enormously, but it somehow has to fit into the confines of the skull. The solution is the same as that required to fit a large area of paper in a fist – you crumple it up.

The principle agents that direct this astonishing process of development are, of course, our genes.

Initially a gene was a hypothetical entity that seemed to be required in order to explain how simple traits, such as colour, could be passed from one generation of sweet peas to the next. Each cell somehow mysteriously had its own set of instructions, its own 'genes', as to what chemicals to manufacture. By the end of the 19th century these still hypothetical instructions were suspected of lurking in the nucleus – the central part of the cell – and perhaps within the chromosomes. Chromosomes are ribbon-like structures that exist in the nucleus in pairs. In humans there are 46 long strands (23 from each parent), but this figure is neither uniform across different life forms nor indicative of the prowess that might eventually be passed to the next generation. For example, goldfish have 94 chromosomes, dogs 78, and cabbage plants 18. Chromosomes are made up of a protein (histone) and a long, chain-like molecule, deoxyribonucleic acid (DNA).

Since 1944 it has been known that DNA is the physical basis of the concept of the gene. Then, in 1953, Francis Crick and James Watson made a ground-breaking discovery, often cited as the most important scientific advance of the 20th century. By working out the molecular structure of DNA they translated the hypothesis of the gene into reality.

Watson and Crick's great discovery was to show how DNA could actually pass on information from one generation to the next. They described the now famous double helix of two entwined strands of DNA, forming a ladder that twisted like a paper streamer. The sides of the ladder were made of sugar and phosphate, but the real excitement arose from how the rungs of the ladder were formed. The rungs were made from two of four possible 'bases' (little units containing nitrogen) – adenine, thymine, cytosine and guanine. The rungs of the ladder are

Deoxyribonucleic acid (DNA) can unzip into two separate strands – the ladder splits down the middle into two matching halves. The isolated strands can then serve as templates for the construction of new DNA, causing the original ladder to double up, or replicate. Alternatively, an isolated strand of DNA can serve as a template for another molecule known as RNA. In this process of manufacturing RNA (transcription), a full-length copy of a gene is made; then this is modified so that only active, coding parts (exons) are present. Finally, the RNA leaves the nucleus and enters the main part of the cell. Since this RNA is to act as an intermediary between a gene and its protein, it is known as messenger RNA (mRNA).

Like DNA, RNA is made up of chemical rungs called bases, of which there are only four types. The task now is to use those four bases as a code for manufacturing proteins. But a protein is a chain of amino acids, and there are 20 different types of amino acid in the human body. How can 4 different bases specify 20 amino acids? The answer is that it takes a sequence three bases (a codon) to code for a single amino acid. The four different bases in RNA can in fact produce 64 different combinations of three – far more than is needed for the 20 amino acids.

mRNA codons do not stick to amino acids directly. This is where another form of RNA, transfer RNA (tRNA), comes in. tRNA acts like the electrical adapter used when travelling abroad to bring together two otherwise incompatible elements. One end of a tRNA matches a particular codon on the mRNA molecule; the other end carries the appropriate amino acid. The tRNA thus enables transfer of a single amino acid to the mRNA, so that amino acids can be brought together to be assembled into the protein of choice, as specified by the gene.

FROM GENE TO PROTEIN

▲ A purified sample of DNA which has been extracted from cells and stored in an alcohol solution.

therefore called base pairs. The DNA of each mammal consists of a total of some 3 billion base pairs, compartmentalized into about 80,000 different segments, or genes; so any one gene will contain tens of thousands of base pairs. If a cell were to divide during development, the exposed bases would pair up with new bases. But adenine will couple only with thymine, and cytosine only with guanine. This means that the new strand of DNA growing on the template provided by a separated strand will be identical to that strand's former partner. Where we had one double helix, we have now grown two; where we had two we now grow four, and so on. This replication of DNA is the actual physical means by which information is passed during cell division.

Genes work by coding for biological compounds called proteins. So if we want to find out how genes affect the brain – not just during development but throughout life – then the key question to ask is how do proteins interact with other brain chemicals and neurons to exert their final effects on the brain? This turns out to be a far from simple matter.

Take, for example, the disease Huntington's chorea, the distressing condition that, as we saw in previous chapters, causes degeneration of neurons in the striatum, resulting in wild, involuntary movements of the limbs. These symptoms do not start to appear until after the age of 40 or so (by which time the individual may have unwittingly passed on the gene to their own children). It turns out that Huntington's chorea is caused by a single faulty gene, which a sufferer may inherit from either parent. Scientists have studied the gene, worked out its DNA sequence, and identified an abnormal protein – huntingtin – made from the gene. But it is still a mystery how this pernicious protein triggers such havoc, and why it takes effect only after age 40. One possibility is that huntingtin interferes with the way energy is used in the neuron; another is that it overloads the neuron with calcium ions, which would swell the cell's power units and kill it. Whatever the reason, the identification of huntingtin shows that even if we can pin a disease down to a single protein, other factors must come into play for its evil to be realized.

Moreover, the triggering of protein manufacture – the activation of the gene – is far from always being a spontaneous or fixed event. Genes inside brain cells can be switched on or off by outside factors, therefore even our 'nature' is far from hard-wired. Just as the division of cells in the developing brain can be initiated within the cell, so their growth and location within the brain are then controlled by the availability of chemicals in the immediate environment. These epigenetic factors play a part almost from the outset, influencing how neurons grow and connect together in the brain of the unborn baby. The prototype substance first came to light over half a century ago.

Rita Levi Montalcini is a remarkable woman. Forbidden from practising medicine by the anti-semitic laws of Mussolini's Italy, she instead set up her own research laboratory. Working in the terrible conditions of wartime and then postwar Europe, she nonetheless contributed to a ground-breaking finding that was to win her the 1986 Nobel Prize for Medicine and Physiology. The breakthrough was the discovery of the first, and most powerful, epigenetic factor known to be involved in brain development – a protein called nerve growth factor (NGF).

NGF has two effects on a neuron: it helps axons and dendrites to grow, and it prevents the neuron from committing suicide. Let's look at the effect on axons and dendrites first. Imagine a growing axon reaching tentatively towards a target cell. If the target cell releases NGF, the protein diffuses across the ever-narrowing gap and enters the tip of the axon. This has the immediate effect of stimulating vigorous growth, encouraging the axon to stretch right up to the target cell and form a synaptic connection. What about the second effect? Some of the NGF in the growing axon journeys back towards the cell body to reach the nucleus – the cell's control centre, and the site in which all the genes are held. Here, NGF triggers a chain of chemical events, the end result of which is that a particular gene in the nucleus is switched on. In ways that are not yet fully understood, the protein made by this gene prevents the neuron from committing suicide – or, to be technically precise, it prevents a process known as genetically programmed cell death – apoptosis.

Apoptosis might sound like an odd idea, but molecular biologists are increasingly realizing that this process is extremely important, not just in humans but in all animals, as it provides a way of getting rid of cells that are no longer needed in the body, even though those cells might be perfectly healthy. Caterpillars, for example, rely on apoptosis to destroy their body cells while they metamorphose into butterflies or moths, and tadpoles lose their tails by apoptosis as they turn into frogs. Apoptosis also turns out to have a role in cancer – cells that fail, for some reason, to obey the chemical commands that trigger apoptosis can proliferate uncontrollably and grow into malignant tumours. Almost any type of cell in the human body may be sacrificed by apoptosis, and brain cells are no exception – hence the importance of NGF.

NGF was just the beginning. A host of other epigenetic factors have since been discovered, among them a class of compounds already known to have a profound effect on the human body – the steroid hormones, such as the sex hormones oestrogen and testosterone. Whereas NGF switches on genes in the nucleus only via an indirect concatenation of chemical events, steroids are far more direct. As soon as they enter a neuron, they combine with special receptors that enable them to penetrate the nucleus itself. Once inside, this receptor–steroid complex clings to, and so temporarily tugs out, a specific gene, causing the cell to translate that gene into a protein. There is good evidence for the involvement of steroid hormones in memory. Oestrogen, for instance, can increase the number connections between neurons, and treatment with this hormone of female patients with Alzheimer's disease has proved beneficial regarding the memory problems associated with that disorder.

As we saw earlier, the neurons of the cortex are produced in their billions in the centre of an unborn baby's brain. During the last few months of pregnancy, they migrate outwards to take up their final positions in the cortex. Epigenetic factors play a key role in this process too, guiding the cells in the right direction. Some are 'chemoattractants' that simply attract the immature neurons to wherever the concentration of the chemical is greatest. Surprisingly, one such chemoattractant turns out to be the neurotransmitter acetylcholine, which can divert the course of neurons growing in a laboratory in under a minute. Others, called adhesion molecules, make the growing neurons stick together or to landmarks in the brain, often in clumps of cells of the same type.

Even before birth, then, the power of the genes can only be realized in the context of a complex and changing chemical environment within the baby's head. And as the brain develops and becomes more elaborate, that environment becomes ever more complex. So what does this tell us about the balance of nature and nurture in shaping our brains? The point is that, although genes may start off and guide the process of brain development, once underway it is open to increasing influence by random effects and hence increasingly varied outside factors. These factors could be as simple as another chemical or as complex as an experience. Nature and nurture, it seems, play an important part in building our brains.

Traditionally, scientists have regarded the unborn baby as unconscious and insentient, lacking a mental life. But more recent evidence suggests that even in the womb experiences can count. Bernard Devlin and colleagues at the University of Pittsburgh School of Medicine analysed the IQ scores of nonidentical and identical twins in order to work out what effect the environment of the womb might have had. Identical twins who are reared apart often have similar IQ scores, a finding that in the past has been attributed to their identical genes. But identical twins have something else in common – a shared experience in the same womb. When Devlin reanalysed the data from nonidentical twins, who have different genes but shared a womb, he found that a significant proportion of the IQ score that had been assumed to be genetic was in fact due to common prenatal experience.

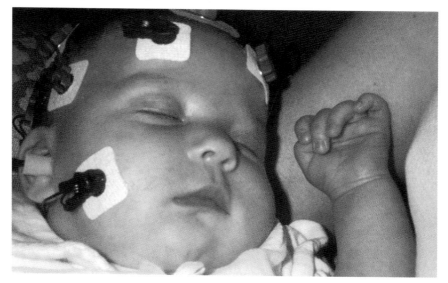

▶ Experiments by Charles Nelson recording the brain activity of infants have shown that memories are already present at birth: a very young baby can differentiate between her mother's voice and that of a stranger.

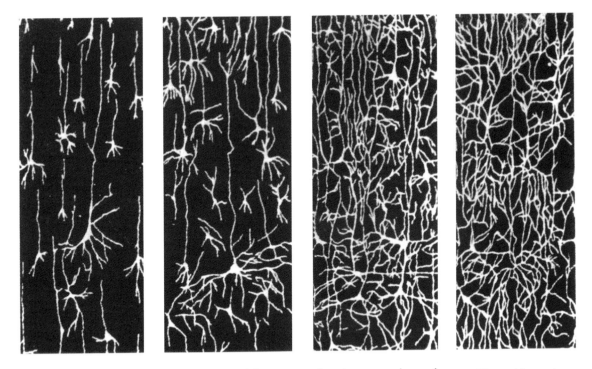

So even at this early stage, external factors are leaving a mark on the brain. More evidence of this comes from the studies of newborn babies by Charles Nelson, Professor of Child Psychology at the Institute of Child Development, University of Minnesota. Nelson has found that memories are already present at birth. He took EEG recordings from newborn babies in response to different voices, and discovered that a baby's brain responds differently to the mother's voice, compared to that of a stranger.

Experiences in the womb pale in comparison with the bombardment of the senses that begins once we are born. How does the brain respond now? The answer lies in the way brain cells connect to each other. Although we are born with almost all the neurons we will ever have – some 100 billion – a great many of the connections between these neurons are forged *after* birth. The growth of new dendrites reaches a peak when we are about 8 months old, and the process continues apace as we grow. So great is this growth of connections, particularly in the cortex, that the brain has quadrupled in size by the time we reach adulthood, despite the number of neurons staying roughly the same.

The advantage of this mind-boggling increase in connections is that it makes the human brain incredibly flexible, able to adapt to the unique life of the individual. Like a muscle, which grows in size and strength if

▲ The rapid growth of connections linking brain cells in the human cortex after birth: (from left to right) at birth, three months, fifteen months and two years.

exercised regularly, the human brain responds to use by making new connections and reforging old ones. And depending on the individual, different parts of the brain may need to adapt more than others.

For example, in blind people who regularly read Braille, there is evidence that a larger area of the brain becomes allocated to the sense of touch. Because such studies must be non-invasive, it is hard to know exactly what these changes might mean in terms of neuron operations. But such questions can be partly answered from observations in rats.

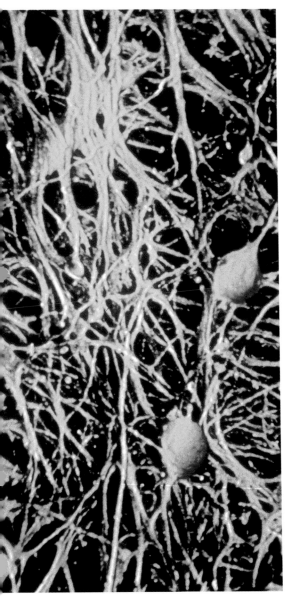

In 1947 the pioneering Canadian psychologist Donald Hebb found that rats kept as pets in his home showed greater learning ability than those kept in a laboratory. Scientists have since repeated Hebb's experiment in more controlled conditions. Rats encouraged to lead a life of Riley, in a cage filled with amusements and the company of other rats, were compared with rats confined to a barren, solitary existence in unstimulating surroundings. After only four days, the Riley-rats have longer dendrites, remodelled synapses and more neuronal connections.

The neurobiologist Dale Purves of Duke University Medical Center, North Carolina, has studied the part of the rat cortex involved in touch, pain and sensation of temperature. This area grows more than the rest of the cortex because, Purves argues, it is used the most. Even within this region some neurons appear to grow more than others. Purves found that the parts of the cortex allocated to the whisker pad and snout, for example, grow to twice the size of the parts allocated to paws. But most exciting of all was his discovery that these enlarged areas use up energy more quickly than the rest of the brain, and have a more extensive blood supply. Purves had found direct evidence that hardworking parts of the brain do indeed grow more connections.

There is one particular, often-used picture of the brain. Taken by a scanning electron microscope, it shows an enormously magnified view of a number of brain cells. A few bulbous cell bodies stand out, but by far the most conspicuous feature is the mass of connections between them – perhaps not surprising when you consider that there can be up to 100,000 dendrites on a single neuron. Because this photograph happens to have a golden tint, it is very arresting. A nonscientist friend, on noticing it in my slide collection, immediately remarked on its beauty, dubbing it 'the Golden Jungle'. The jumble of connections does indeed look the opposite of anything civilized, but the term jungle turns out to be far more appropriate than my friend could have realized.

Just as regular use stimulates the brain to develop more connections, so neglect has the opposite effect. Of course, it would be unethical, not to say impossible, to conduct the kind of deprivation experiments on humans that scientists can carry out on rats. Despite this, Professor Daphne Maurer of McMaster University, Toronto, and psychologist Terri Lewis found an ingenious way to study what happens to the developing human brain when it is deprived of its normal sensory input. Their research reveals the law of the jungle at work in our brains.

Maurer and Lewis examined the vision of children who had been cured of congenital cataract, a condition in which the lens in the eye becomes opaque, blurring vision to such an extent that the affected eye is as good as blind. Most of the children were younger than 7 months old (although some were much older), and, in every case, loss of sight had begun before 6 months of age.

This timing was critical. Humans are not born with the ability to see normally – instead, vision develops gradually over the first year of life as key parts of the cortex develop the connections needed to make sense of nerve signals coming from the eyes. As a result, a child affected by a cataract during this sensitive period will suffer long-term vision impairment, even if the cataract is successfully cured, because the all-important neuron connections have not formed in their brain. As one might expect, the longer the period of sight deprivation, the worse the impairment, and this is exactly what Maurer and Lewis found. But a less predictable finding was the difference between children cured of only one cataract and those cured of two. Curiously, children cured of cataracts in both eyes experienced a greater improvement in vision than those cured of one.

How might we explain this? Let's go back to the jungle, where all the territory is colonized by dense vegetation. When one eye is occluded, the part of the brain that deals with vision fails to develop the connections needed by that eye. However, because the law of the jungle operates, the territory in the brain does not lie vacant – it is taken over by the connections needed by the normal eye. Now, if both eyes are occluded, neither

◄◄ The 'Golden Jungle'. The mass of connections between the brain's bulbous cell bodies can be appreciated in this artificially yellow-tinted, highly magnified electron microscope photograph.

can colonize the visual part of the cortex with their connections, and the cortex retains its adaptability, or 'plasticity', for longer.

This plasticity of the brain in childhood can be remarkable. Harrison is now 6 years old. Soon after he was born, a blood vessel burst in his brain, causing severe epilepsy. His fits were so frequent that it was impossible for him to lead a normal life. In fact, an EEG showed his brain to be in continual turmoil, producing small seizures all the time. A scan revealed that the problem was due to damage in the left side of the brain; the only way to get rid of the storms in Harrison's brain was to remove most of the entire left side. The big danger, however, was that this might cripple him or leave him unable to speak.

Harrison had the operation in June 1999. Happily, when he came round, his fits had disappeared, and he was still able to speak and walk. This miraculous recovery was possible because something extraordinary had happened in Harrison's brain early in his life: years before the operation, his brain had compensated for the damage by allowing the right side to take over the left side's functions. The left side became redundant, and so could be safely removed.

A less dramatic but much more common example of the brain's ability to adapt is 9-year-old Mark's story. Mark was having serious problems reading so he took a course with a specialist reading trainer, Dr Bruce McCandliss of the Center for the Neural Basis of Cognition in Pittsburgh, and his reading skills improved enormously. A brain scan performed before the course revealed that a part of the brain normally involved in reading was underactive in Mark. Remarkably, this region lit up brightly on screen when Mark's brain was scanned after his reading had improved. The issue here is not that there is a part of the brain specifically for reading – there isn't – but that circuits in the brain that are somehow important in this mental skill had been shown to change as a result of learning.

Clearly, then, the human brain is adept at making new connections to suit an individual's unique needs. It is incredibly plastic, able to learn and adapt, to improve and refine whatever skills are most used, purely as a result of stimulation. So how does it do it? How can increased activity in a few neurons actually change their physical shape and the extent of their territory?

A good place to start looking for answers is a nervous system so simple that its cellular nuts and bolts translate directly into behaviour. An unwitting candidate for this lowly plane of existence is a slug-like sea creature called *Aplysia*. *Aplysia* is a 'sea hare' – a kind of aquatic slug that has tentacles rather like a rabbit's ears, hence the name. It cannot do very much, but its behavioural repertoire does include certain reflex actions that are easy to study. Touch its breathing tube, for example, and *Aplysia* quickly withdraws its gills. Although this is a simple reflex, *Aplysia* can learn to modify the response in two distinct ways. When the breathing tube is touched repeatedly, the withdrawal response peters out and the gill eventually stays outside the body. In contrast, if a painful stimulus is applied to *Aplysia*'s head or tail just before the breathing tube is touched, the withdrawal response becomes greatly enhanced.

It turns out that the vigour of *Aplysia*'s gill withdrawal is a direct consequence of the amount of neurotransmitter released from the incoming sensory neuron onto the outgoing neuron that triggers withdrawal. Neuroscientists have accordingly interpreted the two types of learning shown by *Aplysia* in terms of changes in the amount of neurotransmitter released. But there is a problem with this theory. The change in the level of neurotransmitter is temporary, lasting only an hour or so, yet we know that we do not always forget what we have learned after such a short period. There must, therefore, be some additional way of trapping the change so that it persists indefinitely as a memory.

In many species, from *Aplysia* to humans, one way of producing long-lasting changes in a neuron is to switch on certain genes. These might be genes coding for things like receptors, ion channels or neurotransmitters, or genes that affect the cell's shape and connections. One particular trigger that we know can activate such genes is an influx of calcium ions into the cell. So, if calcium is the trigger, the next question is what makes calcium enter the neuron during a learning situation?

A phenomenon that was discovered some 25 years ago – and that still fascinates neuroscientists today – might hold some of the answers. This phenomenon goes by the jaw-breaking name of long-term potentiation (LTP). To demonstrate LTP a cell, most usually in an isolated but living slice of hippocampus, is electrically stimulated with a high frequency current so that it excites a neighbouring neuron. This high-frequency stimulation is repeated over and over again. When an isolated, weak stimulus is applied later – even hours later – the target cell becomes massively excited. In other word, it has become 'potentiated'. At the molecular level, the repeated stimulation has caused a special ion channel to become unplugged, allowing calcium to flood into the cell when the weak stimulus is later applied.

So LTP enables neurons to change the way they act and remember the change for a long period. The big challenge now is for neuroscientists to

▲ The sea slug *Aplysia*, or 'sea hare', has a primitive nervous system which has been much studied by neuro-scientists. *Aplysia* has a simple memory and learns to respond to different types of stimulus in distinct ways, according to the particular neurotrans-mitters released at certain times, under certain conditions.

relate this effect to learning and memory in a whole, living brain. It is one thing to study the effects of electrical stimulation on dissected neurons in a lab, but a different matter entirely to work out how learning works in a living animal. What might be the natural equivalent of the repeated electrical stimulation that causes LTP? One possible clue comes from EEG recordings of animals during periods of activity and rest. While an animal is moving around, its EEG shows rapid bursts of action potentials, but when the animal rests the pattern changes to slow, repetitive brain waves. Perhaps these resting brain waves are the equivalent of the repeated electrical stimulation, serving to consolidate an animal's experiences while it is inactive. Unfortunately, a drawback with this theory is that there is no evidence that animals, especially humans, that rest more frequently have better memories!

Another problem with LTP is that it eventually fades, yet memories last for decades. Perhaps LTP simply sets in train other, unknown processes that lay down our memories permanently. It is important to remember that we are not nano-brained sea slugs. With *Aplysia*, we could progress in one step from a neuronal mechanism to a learned

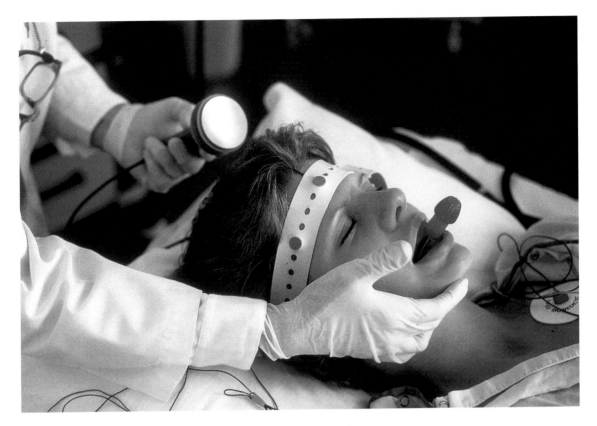

behaviour, but in mammals we are forced to proceed in two steps at the very least, first to mechanisms such as LTP, and then from LTP to an emergent process such as learning. Even if LTP does turn out to be a key part of the learning process, there is no reason to expect its relationship with learning to be exclusive. LTP might well be involved in all sorts of other processes, and likewise, learning may involve many other mechanisms besides LTP. One such mechanism – known as kindling – came to light in a laboratory experiment which was related to the study of epilepsy.

If certain parts of a rat's brain are stimulated for a couple of seconds each day by electrodes implanted in its head, then a strange change in behaviour emerges. For the first few days the stimulation might cause slight facial and chewing movements. By the third day, the same stimulation leads to dramatic jerks of the whole head. By the fifth day, these jerks spread out to the forelimbs. Only four days later, and still with the same level of brief stimulation, the animal is rearing up on its hind legs. Finally, after about 18 days, the treatment causes rearing and falling, and convulsions like those of a full-blown epileptic fit.

Kindling bears some striking similarities to LTP; moreover, we know that both are based on the sustained entry of calcium ions into a target neuron. So, how do they differ? The biggest functional difference is that kindling is more or less permanent. It is immediately therefore an attractive candidate for the biochemical nuts and bolts basis of learning. Moreover, at the cellular level, kindling has been shown not to cause any pathological damage to neurons and indeed it seems to operate over a time scale that would make it complementary to, rather than a rival of, LTP. Kindling continues to develop after the initial induction of enhanced electrical activity in the neurons being recorded, whereas LTP is already at its most potent. LTP can appear within minutes and last a maximum of weeks, whereas kindling takes hours or, more usually, days to develop, but is permanent thereafter. However, the biggest difference between these two mechanisms is not quantity but quality: LTP does not depend on the appearance of convulsions, whereas kindling does. Perhaps such cataclysmic happenings ensure that so much calcium enters the neuron that changes will be permanently reprogrammed. We have also seen that hormones are vital for long-term change; so perhaps it is only with the very powerful, explosive response seen in neurons during kindling that the steroid hormones gain entry into the neuron in significant amounts.

However, even kindling itself, or an exaggerated version of it, can sometimes have opposite effects. In 1933 the Hungarian physician Ladislaus von Meduna introduced the idea that seizures might have a beneficial effect on mental health since there was a very low incidence of schizophrenia and epilepsy in the same patient. Accordingly, he started to give mentally ill patients 'convulsive therapy' by administering the seizure-inducing drug metrazol. This procedure was refined four years later by turning from chemistry to physics and inducing seizure with electric shocks. Treatment with this electroconvulsive therapy (ECT) can also have a marked therapeutic effect on very severe depression when all else, all drug treatments, have failed. Neurotransmitter concentrations that undergo changes during ECT include noradrenaline and serotonin, deficiencies of which have long been associated with depression.

To carry out ECT, the patient is anaesthetized and electrodes are placed on each temple. An electric shock is then passed through these electrodes for about half a second, causing a brief convulsion. The convulsion is essential for treatment to be effective. Typically, the procedure is repeated 1–3 times a week for up to a month. The poet Sylvia Plath, who eventually committed suicide, described the effect of ECT as being like having a bell jar lifted from one's head. Indeed, the effect must have been completely transforming because her tortured quasi-autobiography was titled *The Bell Jar*.

Scientists have found that ECT is most effective when administered in short but repeated sessions, as in kindling. But it is hard to see how the alleviation of depression could be described as learning or memory in the conventional sense. In fact, quite the opposite seems to occur. The most frequent side effect is memory loss, particularly of events in the hour prior to treatment. As with kindling, it seems that 'priming' the neurons is important – memory loss becomes greater with successive treatments.

We have come a long way from looking at an isolated strand of DNA. But it is this nested organization of the brain – from major brain regions to complex neural circuits, to simpler connections, to the single synapse, to the chemicals involved in synaptic transmission, to the genes that produce those chemicals and many others – that makes the nature–nurture distinction so murky. A gene on its own can merely request the formation of a protein; and a lonely protein, outside the context of a complex, working brain, is powerless. Granted, nothing in the brain happens without being reducible, in some sense, to the commands of genes. But then those genes are, in turn, at the command of the epigenetic factors that reflect events outside the neuron.

Even at the most ambitious estimate, the human genome has a mere 1 million (10^6) genes. But the number of connections in the human brain is estimated to be around 100 trillion (10^{14}) – a figure too vast to comprehend – so it is unthinkable that each connection could be programmed by a single gene.

Nature and nurture work along a continuum. In very simple animals, behaviour is more genetically programmed than the product of interaction of the environment. The price paid for this inflexibility is that all the members of a species have a similar behavioural repertoire and lifestyle; it seems to us that goldfish, let alone sea slugs, don't come with a wide range of personalities. In animals with more sophisticated brains, however, more emphasis is placed on learning than acting out the mindless dictates of genes. Cats, for example, have far more personality and individuality than goldfish. And in more complex creatures still, such as humans, that shift from nature to nurture is even greater.

Of course, nature and nurture both play important parts, but they are mutually interactive roles with an increasing emphasis on nurture as we become more individualistic, more human. In order to explore the most exciting question of where the basic roots might lie of individual brains, we now need to explore how our brains interact with the environment.

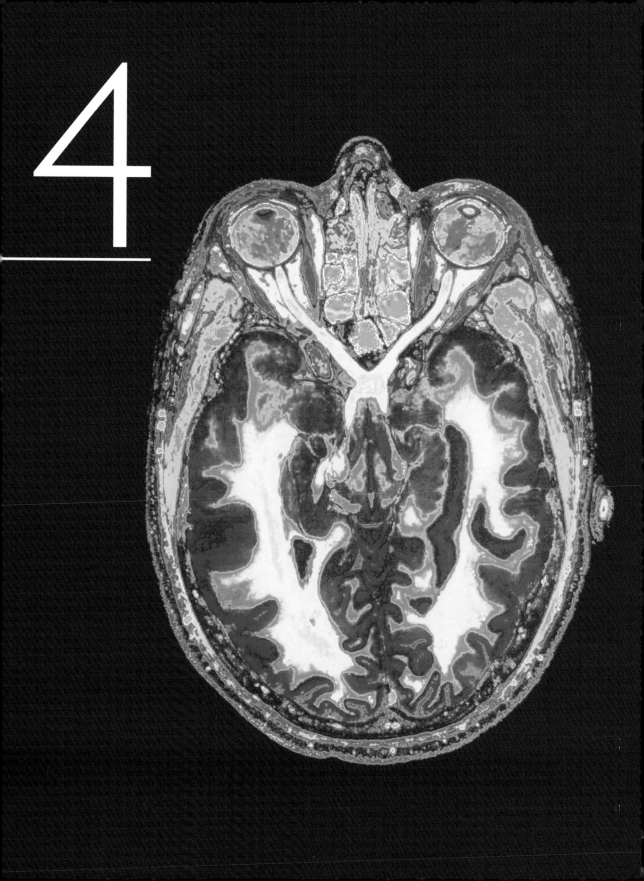

THE MIND'S EYE

Dan Simons at Harvard University has demonstrated a phenomenon that seems unbelievable. Imagine going into a reception area and being greeted by a man at the desk. He asks you to fill in a registration form, which you complete and give back to him; he thanks you, and you take a seat. An everyday experience, perhaps. But if you were then told that the man who took back the form was different from the man who first greeted you, you wouldn't believe it. Surely you would have noticed a change as you looked at the man to hand the form back? Not so – as Dan Simons demonstrated.

He turned this simple scenario into an experiment. While the unsuspecting subject was filling out the bogus form, the man behind the desk changed places with another man who looked completely different, and who had been crouching out of view. Because the subject *expected* that the man would be the same, their brain was oblivious to the very different visual signals coming from their eyes as they returned the form.

We depend on vision more than any other sense to help us negotiate the world about us. Our eyes provide us with a continual stream of information concerning the appearance and movement of people and objects that make up the vast backdrop of our lives *out there*. In this chapter we shall see how *out there* is far from being the reassuringly complete and solid entity that it seems, and how we are far from being passive cerebral sponges. It turns out that we humans, and indeed all other animals, see not with our eyes but with our brains.

A frog placed in a confined space will happily live on the flies that swarm around it. But if dead flies replace living flies, then the frog starves to death. It just cannot see the dead flies, because the frog eye is concerned only with objects that move. In normal frog life, after all, all that matters is whether a looming shadow (a potential predator) or a flitting insect (a potential meal) crosses its field of vision. Frogs don't need fine perception of static objects, so their vision does not cater for this luxury. But the human visual system does. Our sense of sight is far more sophisticated than that of frogs – we can see colours and shapes in great detail, and our eyes detect changes in space (edges) and time (movement)

◀◀ A computer coloured section through the head showing the brain and eyes (at top). The optic nerves (yellow) lead from the eyes (pink) and converge within the brain.

in richer and more diverse contexts. But even we have limitations. Our vision is nowhere near as sharp as that of a hawk, which can spot a tiny rodent scuttling through the undergrowth from hundreds of metres away. Nor can we see objects by ultraviolet light, which might cause a bee to think that our view of the world is unhappily restricted. In each case, eyes and brains have evolved to perform only what the species needs.

How can we find out how our eyes and brains collaborate to create the sense of vision that makes up our own version of reality? We need to start by taking a brief lesson in anatomy and finding out more about the special neurons that carry the signals from our eyes deep into the brain.

At the back of our eyes lies the retina, from *retus* – 'net' in Latin – because it is made up of a dense network of cells and connections. The retina is like a screen – light entering the eye is focused by a transparent lens just behind the pupil and projected onto the retina, where it forms a crisp image. This image is detected by special types of neuron in the retina that are sensitive to light. When stimulated, these produce electrical signals that are relayed to neurons running into the brain. The connecting neurons carry the signals along the two optic nerves (one from each eye) to a special structure on each side of the brain called the lateral geniculate nucleus (LGN), from where the signals are forwarded to the visual cortex, the part of the cortex concerned with vision.

As there are two optic nerves and two LGNs, it might seem that each LGN processes the signals from just one eye. However, it is more complicated than that. If one of the LGNs becomes damaged, by a stroke for example, the person becomes blind not in one eye but in half of each eye. This is because of the way neurons in the retina connect to the LGN. Neurons from the left side of each retina connect to the LGN on the left side of the head; neurons from the right side of each retina connect to the right LGN. So, inputs from both eyes are represented in both LGNs.

The picture gets more complicated still. Scientists recently discovered that the whole visual system really consists of two systems working in parallel. There are actually two distinct types of cell – large and small – that relay signals from the retina to the brain. For the sake of simplicity let's just think of the visual pathways as simply divvied up into two parallel systems, system A (large cell) and system B (small cell). This A/B distinction holds true in the LGN – certain layers of each LGN are made up of cells of the A system, and other layers are made up of cells of the B system. Next, neurons run from the LGN to the visual cortex. This is divided into a sequence of regions, V1 to V6, that again are divided into the A and B systems.

How might we study how vision is processed in the brain? The basic procedure is to present an image to one or both eyes and then record any changes in activity, as measured by the frequency of action potentials, in key parts of the brain, such as the LGN and the different parts of the visual cortex. The Nobel Prize-winning scientists David Hubel and Torsten Weisel of Harvard Medical School pioneered this strategy. They used electrodes to monitor electrical activity in cats' brains in response to flashes of light.

Hubel and Weisel found that the cells of the LGN gave a simple on/off response to a spot or ring of light. But the response of cells in the visual cortex became increasingly fussy as they got further from the LGN. The cells of V1, each of which receives input from several LGN cells, also responded to a spot of light, but they were even more sensitive to rectangular bars of light at particular angles. The cells of V2 were not excited by spots at all, only by bars at certain angles. Finally, the cells beyond V1 and V2 responded only to bars of a particular length.

It turns out that the A and B systems take on different aspects of the job of seeing. The A system is particularly involved with movement. It might not feel as if we process motion separately from everything else that we see in the world, but in fact this is the case.

In 1981, German Professor Joseph Zihl was introduced to a patient – Gisela Leibold – who was confounding her psychiatrists. Frau Leibold had recently suffered a stroke that affected a specific part of her visual cortex. Although she had seemingly recovered well, she now suffered chronic agoraphobia. To Zihl's astonishment, it turned out that Frau Leibold's problem was neurological, not psychiatric. As a result of damage to the part of her visual system concerned with movement, she can see perfectly well but is blind to movement. For instance, if shown a picture of a boy falling off a chair, she can predict that he will be hurt. But when she goes outdoors she finds herself in a world where cars are almost on top of her, and where she can take no avoiding action from what is happening around her. It is, understandably, an unsettling and disorientating experience.

'I find it very disturbing. When the fast underground train comes, it is not pleasant for me', she says. 'If it is coming towards me, it is worse. It is better when it is far away. I try not to look at anything that is moving towards me, that is moving directly in front of me.'

To make sense of what is going on in the world, Frau Leibold waits for the static image in her mind to update. After a delay of a few seconds, she can deduce that something is moving from the change in this snapshot.

▲ Gisela Leibold. As a result of a stroke, she cannot see movement, although her vision in other respects is completely unimpaired.

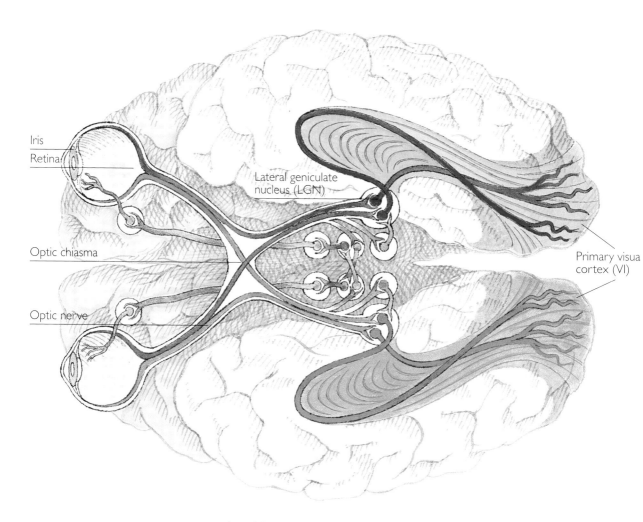

Iris
Retina

Lateral geniculate
nucleus (LGN)

Optic chiasma

Primary visua
cortex (VI)

Optic nerve

▲ The sight systems of the brain. Light entering the eyes is focused on the retinas where special cells convert it into electrical impulses. These are carried into the brain via the optic nerves to the optic chiasma, and then to the lateral geniculate nucleus (LGN) and the primary visual cortex. In the primary visual cortex, 'VI', there is some initial sorting before sending the data to the association visual cortex around it.

Joseph Zihl: 'It is very difficult to imagine how the world of Frau Leibold may look like. The problem is that she cannot see movement in the sense that tells us whether something is moving or not, or tells us the direction of or the speed of movement. Moving objects appear to her as restless, and this is what makes her feel very unwell and irritates her and makes life so difficult for her. So, it's a mixture of having snapshots of the world if something is moving, and at the same time having this strange impression of restlessness'.

Movement and detection of movement are vital parts of our lives, but compared to many other animals we are fairly slow-moving and our ability to see motion is limited. The illusion of the cinema reveals just how easy it is to trick the human brain into believing a still image is moving – 24 frames a second is good enough for us, but for many animals it would be like watching the very earliest flickering movies.

If the A system deals with movement, what is happening in the B system? The B system divides into two sub-systems, one concerned mainly with colour and the other with form. Again, evidence for the remarkable way in which the brain divvies up the job comes from patients with highly selective brain damage.

People with damage to certain parts of the V4 area suffer from an extreme form of colour blindness known as monochromatism, in which the world is seen in shades of grey or beige. Accordingly, when neurologist Oliver Sachs asked one such patient to paint a collection of fruit, the orange, banana and apple all came out the same muddy and uniform colour. People with damage to only one side of V4 may have the bizarre experience of seeing half the world in colour and half in monochrome.

Finally, there are those who have suffered brain damage outside the visual cortex but who still have problems with their sight. One such

▲ A drawing from memory by a patient suffering from achromatopsia, a type of colour blindness where the world appears in monochrome. The pictures show how well his form vision is preserved but how deeply colour vision has been affected. The banana, tomato, cantaloupe and leaves (clockwise from top left) are all depicted in uniform hues.

condition is known as neglect. It leaves certain stroke patients with a baffling mixture of symptoms as the damaged brain recovers – it can seem as if half their world has disappeared.

The stroke that Peggy Palmer suffered 10 years ago left her with extreme neglect on the left side of her world. Just as a magician draws your attention to one side so that you don't notice the sleight of hand on the other, so Peggy's attention is so skewed to the right that she seems to see nothing at all on the left. The damage to Peggy's brain occurred in the parietal cortex, yet what she sees is nonetheless profoundly impaired.

Her neuropsychologist, Peter Halligan, explains: 'neglect has very little to do with your eyes. The vast majority of the problem arises from the brain processes involved in attention. Your attentional system provides for where your eyes move. So, in other words, if something happens in my visual field that is interesting, I'll move my eyes there. But why would you move your eyes there? Only if your attentional system indicated the need to move there. So your eyes are slaves to your attentional system – and what's wrong in neglect is the attentional system has been damaged.' Peggy's brain is filling in the gaps, completing the parts that are missing. This means that a person with neglect almost never notices the mistakes they make until their attention is drawn to it.

Peggy tries to describe the disorientation: 'It's a bit difficult walking around. When I go down to the day centre, for instance, they take me into the dining room to have lunch. And they will insist on me walking in from the left. And I can't do it because my head will just not let me go that side. It just won't – it's very strange – it won't let me go there.'

Although many parts of the brain seem to be involved, in parallel, in making different contributions to vision, we cannot necessarily say that these brain regions are working independently. Interestingly enough, patients in whom the B system is completely destroyed do not lose the ability to see form, even though form and colour are both processed in the B system. The reason is believed to be that the A system is also involved in processing form, though not colour. If the B system is destroyed, the functional loss is partially compensated by the A system. So although different aspects of vision are processed separately, the relevant parts of the brain do not have exclusive control of any one aspect.

At all levels of visual processing – in the retina, the LGN and the visual cortex – there are connections between the A and B systems. Likewise, there are links between the two ultimate target areas of the visual cortex, namely the parietal and inferotemporal cortices. So although the visual system is made up of two clear-cut systems, these act

▼ Peggy Palmer, who suffers from a condition known as neglect: she can only see on the right side; on her left she is in effect blind.

more like a married couple than two strangers. How, then, do all these tidy, compartmentalized brain regions, with their interconnections and subsystems, actually work together? What do they actually *do*?

One of the earliest schemes proposed – inspired by Hubel and Weisel's finding that neurons involved in vision become increasingly fussy as they get further from the retina – was of a kind of hierarchy. The idea was that, in the end, fewer and fewer cells would respond to increasingly complex patterns, in a pyramid-type arrangement. Eventually, we might imagine a single cell responding to a complex image of, say, your grandmother. But could this process alone, leading to a single 'grandmother cell', account for the recognition of the same grandmother in all her different moods, replete with all her outfits? And is there any evidence that certain neurons are responsible for such a specific task as recognizing faces?

Thirty years ago Lincoln Holmes was in a car accident that almost cost him his life. Slowly, he recovered from his injuries, but he was left with a very unusual handicap: although his vision was normal in all other respects, he could no longer recognize faces. Lincoln was a victim of a prosopagnosia, a condition caused by damage to the inferotemporal cortex. He sums up the condition with a revealing anecdote: 'some years ago I was going to a conference in Boston and at some point I left the auditorium where the conference was being held to use the lavatory a.k.a. washroom. On my way back, I came round the corner and saw what I thought was a friend of mine, and greeted him by name. There being no response, I looked again and realized that I was looking at a mirrored wall, and therefore at myself.'

Lincoln is learning to use other cues to help him recognize a face – gender, height, voice and context. But how do the rest of us achieve such a sophisticated process so effortlessly? One idea is that the inferotemporal cortex might contain groups of highly specific recognition cells, or 'gnostic units', for recognizing faces. In a sense, then, such units would come at the top of a hierarchy of increasingly fussy cells. However, there are two main objections to this idea. Firstly, the idea of small groups of neurons equipped to work as independent mini-brains is hard to accept – such units could not function without input from other complex processes, such as memory. And secondly, there is a limited number of neurons in the brain. Would we cease to recognize objects after our supply of gnostic units had been exhausted? If a gnostic unit were to succumb at random to the daily neuronal death toll, would we suddenly become unable to recognize our grandmother? Such experiences are never reported.

▼Lincoln Holmes cannot recognize faces, a condition known as prosopagnosia.

But the biggest problem with the hierarchical theory is what happens at the end. To whom does the final cell report? We can get round this problem to some extent if we think of the parts of the visual system working in parallel. We have already seen that different aspects of vision can be processed independently, but that at all stages there is interaction and feedback rather than complete independence, just as different members of a football team specialize in attack or defence but nonetheless work together. This type of interaction might form the basis of recognition of faces and other complex visual scenes, and provide an alternative scheme to the unlikely 'grandmother cell' scenario.

The German physiologist Wolf Singer has recorded the activity of many neurons simultaneously when the eyes see a particular stimulus. Singer found that neurons in the visual cortex of a cat, sometimes as far apart as 7 mm (some 500 times wider than a neuron), become active in synchrony when a particular pattern is presented. He has shown that such distant neurons have a transient, synchronized action, rather than one that is long-lasting. So perhaps neurons in the visual cortex form transient groups for processing highly specific patterns under particular circumstances. These assemblies of neurons would differ from the hypothetical gnostic units, not only because the extent of networking is much larger, but also because it is much more dynamic. In this scenario, we need not worry about running out of neurons. If different assemblies of neurons could form and reform to recognize different patterns, we would have a virtually infinite number of combinations. But what might be the deciding factor that recruits the neurons into a temporary group? A possible clue comes from recent studies on the developing visual system.

For vision to develop normally in children, both eyes have to be used during a 'sensitive period' in early life. Only then can the neurons of the visual cortex form the connections needed to interpret signals coming from the eyes correctly. This process of fine-tuning in the visual cortex requires that the eyes focus their attention on particular objects so that the cortex can learn what those objects look like. Clearly, then, arousal – how attentive and alert you are – might play a key role in the development of sight. This theory is supported by experiments in the lab. If a part of the brain called the reticular nucleus – which is involved in arousal but is not part of the visual system – is destroyed in an animal, then the fine-tuning of the visual cortex fails to occur and vision does not develop properly. But, amazingly, that fine-tuning can still occur if electrodes are implanted into the animal's brain to stimulate its arousal level artificially.

So what is going on at the nuts-and-bolts level of molecules and neurons? During states of high arousal, the neurotransmitter acetyl-choline is released into the cortex by neurons running from the basal forebrain and brainstem. Now, it turns out that acetylcholine can act not just as a conventional neurotransmitter – it can also put cells in a state of 'red alert', primed for incorporation into a temporary network. Once they are bathed in a fountain of acetylcholine, neurons that do not normally excite each other are more easily recruited into a working group. During arousal, then, the chance of connections forming between neurons increases. In development, such changes would be more or less permanent and could form the basis of visual pattern recognition.

As we grow, we end up not needing to pay as much attention to what is *out there* in the real world. US psychologist Stephen Kosslyn is fasci-nated by what happens in the brain when we experience a visual scene that is quite simply not out there, namely when we using not our eyes, but our imagination. Kosslyn has spent 30 years investigating whether we use the same parts of the brain when we are imagining something as when we are looking at the world.

Kosslyn's theory is simple: he thinks we use the same areas for vision and for our visual imagination. In effect, he is suggesting that imagina-tion is vision running backwards. Using brain scans, he has been able to show that when we imagine an object, it is indeed relayed all the way back to the early parts of the visual system. But most important of all, this 'backward projection' might help explain how we see normally. Kosslyn's idea is that the backward projection process that operates in our imagination is also a powerful mechanism for our normal perception of the world. Scientists studying vision have known for a long time that the visual system is far more than a one-way street from retina to cortex. We now know that for every connection carrying information from the eyes, there are *at least ten* coming in exactly the opposite direction from the higher areas of the brain. It seems that the information leaving our retina is not complete enough to create a full and rich interpretation of the world. Our imagination, then, allows us to fill in the gaps and con-vert the distorted image from the eyes into the complete and vibrant world that we see.

So even normal vision might lie at the interface between vision and imagination. Think about looking at the moon and seeing a face – the man in the moon. The illusion arises because the information entering your eyes consists of a bright, round object with shadows; the back-pro-jection system fills in the rest, giving the most likely scenario – in this case a face.

However, as often happens in science, just after Kosslyn seemed to have established his theory, new data came along that made him think again. In this case, the data came from a patient, Kevin. In 1988 Kevin had a traffic accident that left him with a serious disability, even though no obvious damage was visible on his brain scans. Whereas most of us are at pains even to see the man in the moon, Kevin sees faces everywhere and in everything, because his ability to imagine faces is overactive. On the other hand, he cannot recognize objects, a not uncommon condition known as object agnosia.

Agnosia is a more complex disorder than the impairments in perception of colour, form or motion that we touched on earlier. It is a problem not with what we see, but rather with the meaning we attach to what we see. Kevin has no basic visual defects – he can see flashes of light, fine grating patterns, and so on. He just doesn't recognize what he sees. When shown a picture of an apple or an open book, he fails to recognize them as such. And despite having no problem with hand–eye coordination, he cannot copy such pictures with any degree of success. On the other hand, if asked to draw an apple or a book from memory, he can do so perfectly well. But if he sees his own drawings later, he does not recognize what they are.

Kevin's ability to recall objects from memory – his imagination – can therefore be dissociated from the process of relaying information from retina to brain. Imagination and normal vision would seem, therefore, to be different processes – the opposite to Stephen Kosslyn's theory. Kosslyn's response to this apparent contradiction has been to use brain scans to show that there is neither complete overlap of the brain regions involved nor complete segregation.

It certainly seems a plausible and attractive idea that our visual experience, in normal circumstances, is a kind of mixture of information coming in from the eyes and prior associations – how else might we interpret what we see and give the world significance? In the case of small children, one can imagine the interaction between brain and outside world being very one-sided, simply because a child has little experience to add to the riot of abstract, literally meaningless shapes and colours pelting its visual cortex. There is another, more familiar situation where the dialogue is one-sided, but this time in the opposite direction. It is a little like using our imagination to daydream, but this time the block on the retinal input is complete – it is when we are dreaming.

Dreaming has, of course, been a source of fascination to mankind since the dawn of time. Nowadays many people would go along with the idea that it is a form of consciousness, easily distinguishable from dream-less sleep. Ever since the introduction of the EEG machine in the 1920s – which made it possible to monitor brain waves during sleep via electrodes on the scalp – we have known that the EEG pattern of sleep is very different from that of wakefulness. The exception is during dreaming, when the pattern is indistinguishable from when we are awake.

More recently, Rodolfo Llinas, Professor of Physiology and Neuroscience at New York University School of Medicine, has used modern technology to show with still more precision just how dreaming and wakefulness, in brain terms, are so similar. Llinas, together with his colleagues, has shown that in both wakefulness and dreaming, but *not* in ordinary sleep, neurons in the thalamus and cortex are synchronized, with both areas generating rhythmic waves of electrical signals in step with each other. All that can be said of this study at present is that, yes, dreaming is a form of consciousness, and that for at least one basic factor – the concerted dialogue of two brain regions – input from the retina is irrelevant. Llinas's own interpretation goes further than this though. He suggests that our brains are in a constant state of dreaming – that they are continually generating images to manufacture the world inside our heads.

'The outside world is a projection, you put it there', says Llinas. 'It is not happening out there, it is happening inside your head. It is, in fact, a dream, exactly like when you fall asleep. We need to see, we need to perceive, we need to dream actively – because this is the only way we can take this huge universe and put it inside a very tiny head. We fold it, we make an image, and then we project it out.'

Positron emission tomography (PET) has also been used to explore what happens in the brain during dreaming. Pierre Maquet and his colleagues in Belgium have shown that there is, after all, a difference in the pattern of activity during dreaming and wakefulness. Parts of the brain involved in arousal and emotion light up in just the same way during dreams as when we are awake. However, large expanses of cortex are far less active. Interestingly enough, one of these cortical areas, the prefrontal cortex, is also underactive in schizophrenia. Schizophrenia has often been compared to dreams, in that both states are characterized by strong emotion yet a paucity of logic or reasoning ability. Could the prefrontal cortex be the key area for differentiating the inner world of our imagination and dreams from the outer world of reality?

Although the cortex is traditionally regarded as the most advanced part of the brain – the site of 'higher' functions like logic and reasoning –

we saw in chapter one that allocating specific brain regions to specific functions was simplistic. On the other hand, the prefrontal cortex is twice the size it should be for a primate of our body weight, and damage here does tend to lead to problems in more sophisticated types of thinking. In particular, damage to the prefrontal cortex can result in 'source amnesia', in which the patient has a memory but cannot place it in a specific frame of reference. In a similar way, we find it hard in dreams to ascribe a precise location or time to an experience – instead, dreams resemble more the experiences of early childhood. Perhaps this type of state, in which one has a kind of passive experience without engaging in complex thought – is what takes place in a dream, the brain freewheeling without restraint from the prefrontal cortex.

The issue of the nature of dreams raises more questions than answers, and we are still left wondering how the brain enables us to be aware of what we see. In our search for answers, we are forced to look for clues in other places, and one such clue could come from a fascinating clinical condition known as blindsight.

In 1917, when World War I was generating a host of patients with brain injuries, a British army doctor called George Riddoch reported cases of people who were blind in parts of the visual field, but who could nonetheless detect motion or experience a vague awareness of objects they could not actually see. These reports were too fantastic to be

▼ Sleep research. A woman wired to somniography equipment. Electrodes attached to the scalp record electrical activity in different parts of the brain.
▼▶ A vivid impression of dreaming – *The Dream*, a painting by Franz Marc (1880–1915).

accepted by the establishment, and it was not until US psychologist Larry Weiskrantz reported similar cases in victims of head injuries some 50 years later that the syndrome of blindsight was finally acknowledged.

In blindsight, patients report seeing 'something' that they cannot identify, or they simply have a 'feeling' that something is there. They can see, yet they are not consciously aware of what they see. Because of this, blindsight has often been seen as the ultimate litmus test for exploring conscious perception, allowing us to cross the Rubicon and investigate the mystery of consciousness. As yet, there is no clear consensus as to how blindsight works. One idea is that the first part of the visual cortex, V1, becomes damaged, but the other parts of the visual system continue operating without it. Another idea, advanced by the vision expert Semir Zeki, co-head of the Wellcome Department of Cognitive Neurology at University College London, is that the connections between V1 and V5, the area vital for motion detection, become damaged.

The field of blindsight research owes much to a man called Graham Young. Graham can see, but he is unaware of anything on his left-hand side. He has lost consciousness of only part of his visual field, an impairment that has made him a prime subject for studies of blindsight.

Graham describes what happened to him: 'I had a road accident when I was eight, resulting in some brain damage ... I lost all my vision to my right in both eyes. I literally used to be walking in town as an 8- or 9-year-old and walk into a lamppost. I'm completely unaware of an object moving in my blind field.'

In Graham's case, what is lost is part of the pathway running to the prefrontal cortex. But what is particularly interesting is that Graham is still processing information subconsciously. In one experiment he was unaware of a light on the left-hand side of a screen, yet when asked to guess where it was he answered with 100% accuracy. Somehow, he is processing the information covertly by bypassing the cortex and using instead one of the primitive pathways below it. Of course, this does not prove that the cortex is the centre for consciousness, but rather that certain pathways between the cortex and the rest of the brain have to be intact.

Again, brain scans have provided a way of investigating the phenomenon. It turns out that there is a difference in the scans produced when Graham is consciously seeing and when he is not consciously seeing. When conscious of an image, and only then, Graham's brain shows up widespread activity – not just in the visual cortex but also in more sophisticated brain regions, such as the prefrontal cortex.

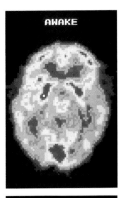

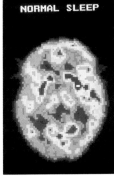

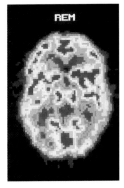

▲ PET scans taken when awake and during normal and REM sleep. Active brain areas are shown in red and less active areas in blue. During REM sleep, the brain is dreaming and shows similar activity to the awake state. In normal sleep, the brain is less active than when awake.

Strangely, Graham has recently started to experience novel sensations in his blind field. He is becoming aware of certain things, such as moving objects, but it is not a normal seeing experience but more of a feeling – Graham describes it as 'black on black'. Could this mean that the primitive subcortical pathways are starting to acquire the ability, somehow, to generate awareness? The simplest explanation is that Graham's primitive visual pathways, ever plastic as we saw in the last chapter, have created new connections to the cortex, thereby creating a type of conscious awareness. Perhaps the primitive 'black on black' awareness is all these circuits are capable of – not sufficiently complex to appreciate the Mona Lisa, but good enough to give warning of approaching danger, such as a fast-moving predator.

The data garnered from Graham Young suggest that conscious vision involves several – probably many – interacting brain areas, of which the prefrontal cortex and the visual system (in its widest reaches to infero-temporal and parietal cortices) are just a few. This idea is borne out by recent reports from neurobiologist Gerald Edelman, of the Institute of Neuroscience, San Diego, California, who has assembled a veritable squadron of researchers to try and find the difference between passive visual processing and conscious vision.

Edelman used an advanced version of EEG technology called MEG scanning. The MEG scanner can detect tiny areas of electrical activity produced in the cortex over incredibly short timescales. The experiment is simple: all a person has to do is place their head in the scanner helmet and wear an unusual pair of glasses. Each lens is different – one has vertical red stripes flashing seven times a second, the other has horizontal green stripes flashing slightly faster. Oddly, it is only ever possible to see one display at a time. It is rather like looking at that well-known image of a vase made of two confronting profiles – one's conscious perception incessantly flicks from one to the other. And this alternation shows up on the brain scan.

Edelman found that when the subject reports consciousness of a certain display, constellations of neurons produce waves of electrical activity at the same frequency as the flashing stripes. And these active areas are seen not only in the prefrontal cortex, the part of the brain traditionally associated with consciousness. They occur in a huge, conspicuous network across the whole brain. Once more, it seems that many parts of the brain are pressed into service during a mental process – and understanding that process is not simply a matter of identifying the parts of the brain involved.

To complicate matters further, we have to remember that we do not experience the world as a purely visual phenomenon. Although vision is our primary sense, our conscious experience is, of course, multisensory. Normally, we can dissect this global experience into sight, sound, touch, smell and taste without difficulty. But a minority of individuals report sensation in one modality when actually experiencing another. Such people, for example, can hear colours or see sounds. This bizarre phenomenon, called synaesthesia, as yet has no convincing explanation. But perhaps it is not so surprising. After all, neurons behave in the same way in all the parts of the brain concerned with the senses – an action potential in the visual cortex is exactly the same as an action potential in the auditory cortex.

The neurologist Richard Cytowic thinks synaesthesia may be a primitive sensory mode in which early animals perceived the world. But, even if true, such an idea does not explain the differentiation of which we are now capable, nor how that differentiation can break down again into synaesthesia. Another possibility is that there are connections between each of the relevant brain areas, and that signals intended for one part are sometimes transmitted to others in error. Yet another suggestion is that synaesthesia is an issue of overactive associations. In one case, a woman always experienced the taste of baked beans when she heard the name 'Francis'. Such a selective pairing seems to hint at a particular, covert connection between memory and, in this case, hearing. A more controlled example of this pairing effect comes from an experiment in which a subject wore special glasses for 20 days. One eye saw the world through a blue lens, the other through a yellow lens. When the glasses were removed, the subject still reported a distortion in vision. When he turned his neck one way he saw the colour complement of blue (yellow), and vice versa. The interpretation runs that he had started to associate movements in his neck muscles with an ensuing experience of colour.

Although these theories have yet to be confirmed, the case of synaesthesia, like many of the examples we have discussed, points to the fact that there is far more to the senses than the brain acting as a mere sponge to the flood of light, sound, taste, smell and touch sensations coming from the outside world. We have seen that we are denied a full view of the world, and of what we do see, much is supplied from within, back-projected from higher parts of the brain. We see things with our brains, not our eyes. If reality is indeed not *out there* but inside our heads, then what we see must depend on the unique contents of our personalized brains – our memory. And that, accordingly, is the subject of the next chapter.

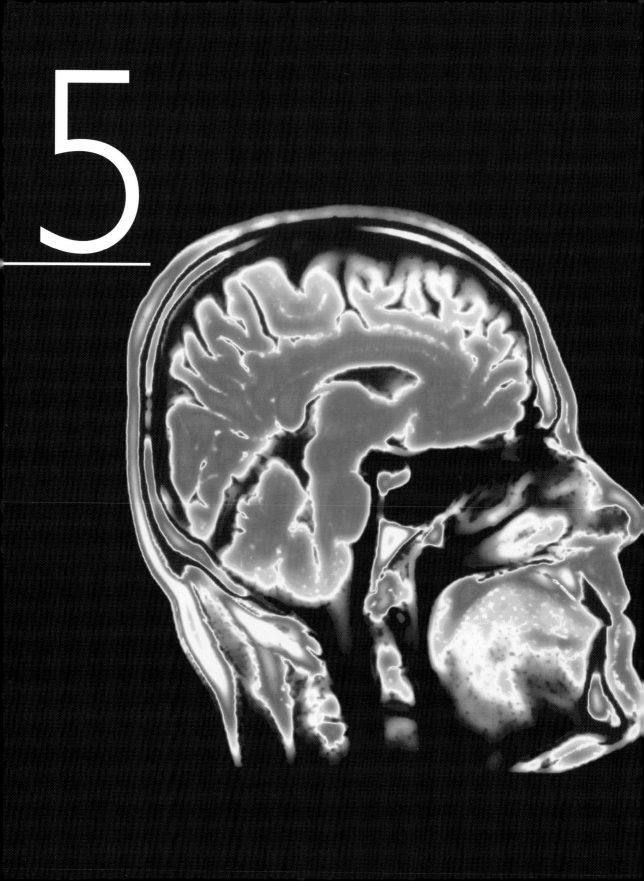

5

THE MEMORY BANK

Harold (not his real name) had severe epilepsy. He was in his twenties when, in 1953, the only solution to the debilitating series of daily fits lay in surgery. Large areas of both sides of his brain were removed. The procedure did indeed successfully combat the fits, but with the terrible consequence that Harold could no longer remember anything that happened to him.

Some aspects of Harold's memory were fine though. For instance, he could still learn how to trace a star-shaped pattern while looking in a mirror – although he couldn't remember the actual event, the training session, itself. As well as destroying his ability to form new memories, the operation on Harold's brain erased his memories of the two years prior to surgery, though not any earlier. Somehow, therefore, the areas of brain that had been removed were playing a part in laying down new memories, memories less than two years old, but thereafter these key brain regions were not apparently needed. So we could hardly think of the extirpated brain tissue as holding Harold's 'memory bank', the database of his lifetime's memories. Instead, this now famous case suggests that memory must be organized in a very different way from a simple store of data. In this chapter we shall explore just how much is known about that organization.

First, we need to be clear what me mean by the word memory. For most of us, a memory involves a conscious thought about a one-off fact or experience. From this point of view, Harold's 'memory' of the ability to trace stars in a mirror is not really an example of memory, but a learnt skill, akin to learning to drive a car or play the piano.

We can divide our conscious memory of facts and experiences into two types: short-term and long-term. Short-term memory requires active concentration, such as when one is trying to memorize a phone number before writing it down, and thus it seems to be as much about attention as memory. Again, Harold's short-term memory was fine – he had no difficulty reciting a string of numbers shortly after reading them.

Long-term memory is perhaps what we think of conventional memory – the ability to deliberately recall snippets of information or experiences from our past. Again, we can divide this type of memory into two different categories – episodic and factual. Episodic memory is our

◄◄ A colour computed tomography (CT) scan of a healthy brain seen from the side.

memory of experiences and events (episodes), a particular birthday party, for instance. Factual memory (semantic memory) is simply memory for facts – the meanings of words, people's names, the identities of tools, and so on. The critical difference between episodic and factual memory is that the former relies on time and location as a frame of reference, the latter does not. Your knowledge of the names of Renaissance painters, for instance, floats free of any particular event in your life, but your last birthday party is probably rooted firmly in a particular place and time.

Let's go back to Harold. The parts of his brain that were removed were the medial temporal lobes, which are situated, as their name suggests, near the temples. The mass of tissue taken out was really quite substantial in that it included several different areas (the hippocampus, entorhinal cortex, medial temporal cortex and temporal stem). From clinical studies such as Harold's, this conglomerate of neurons must clearly have something important to do with memory.

Further evidence that the medial temporal lobes play a role in memory can be seen when nature, sadly, makes lesions all on her own. British neurologist Kim Jobst and pharmacologist David Smith of Oxford University recently gathered evidence showing that as Alzheimer's disease progresses – causing attendant loss of memory – the medial temporal lobes gradually shrink. Jobst and Smith compared this decline in size with the natural shrinkage that occurs in healthy people of a similar age, and found the downward gradient was much sharper in patients with Alzheimer's, alarmingly so. The data showed conclusively that Alzheimer's disease, contrary to what many fear, is not a natural and inevitable consequence of ageing.

Although no one would deny that the medial temporal lobe obviously plays a role in memory, the big challenge now is to find out exactly what that role might be. One problem is that, as we have seen, the medial temporal lobe is really an anatomical collective encompassing a variety of different brain regions that might each have separate contributions to make. Evidence is coming to light that this is indeed the case.

There is a degenerative condition that is sometimes confused with Alzheimer's disease, but which is actually quite different, called Pick's disease. We met this disease briefly in chapter one with the musician Dick Lingham and his new-found painting skills. Pick's disease is less well-known than Alzheimer's because it is less frequent – about one case is reported for every 50 of Alzheimer's disease. Although both diseases cause dementia in later life, there are key differences that provide a valuable insight into how memory is processed in different parts of the

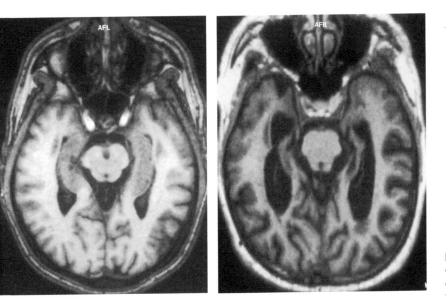

◀ A normal temporal lobe (far left) compared with that of a patient with Alzheimer's disease. The loss of the medial temporal lobe (in this figure seen as the semi-circles surrounding the central part of the brain) has led to enlarged fluid-filled cavities (ventricles), seen here as dark spaces.

medial temporal lobe. The first sign of Alzheimer's disease is memory impairment, whereas the first sign of Pick's disease is a change in personality and behaviour, as we saw with Dick. The patient might start to show speech impairments, such as scanty speech, verbal circumlocutions or an inability to name objects. These early symptoms relate to the initial site of damage in the brain. In Alzheimer's disease, the problem starts from the inner structures of the brain, eventually spreading to the medial temporal lobes and affecting the cortex. In Pick's disease, however, the opposite occurs: damage starts in the cortex, and from the outset it affects a key part of each medial temporal lobe called the temporal cortex.

Psychologist Karalyn Patterson, of the Medical Research Council's Cognition and Brain Science Unit, Cambridge, and neurologist John Hodges, of Addenbrooke's Hospital, believe the symptoms of Pick's disease can be interpreted primarily as a problem with factual memory. If so, this selective impairment in recalling facts rather than experience might explain why patients seem to undergo such a radical change. After all, imagine what it would be like if you no longer knew what a giraffe was – your whole database would be starting to fail you. Inevitably, your behaviour would have to change also, and with it your apparent character.

With Karalyn's colleague Matt I underwent some of the tests used to assess patients suspected of having Pick's disease. They were alarmingly basic. First I had to name pictures on cards: a cat, a spade, a bottle; obviously, this test was accessing my factual memory. But perhaps someone might recognize the objects well enough and simply have a problem with the words. In order to test for a defect in factual memory without the need for words, Matt then produced a series of objects. He told me to hold them as though I was going to use them, and mime or describe the task to which the objects were suited. A paperclip followed a bottle-opener, and then a toothbrush and a hairbrush. There was a garden trowel, a wallpaper stripper and a washing-up brush. As I went through the motions of brushing my teeth and opening a bottle of beer I marvelled at the problems that someone might have if they no longer recognized what these objects were for. But why had Matt included three different types of brush? That was the whole point. As factual memory starts to break up, it does so gradually, causing our compartmentalized knowledge of the world to crumble. Initially, the compartments just get bigger and more general. So someone with Pick's disease might attempt to clean their teeth with the washing-up brush; they recognize objects belonging to the category 'brush', but they cannot discriminate between different types.

The same deficit is revealed in the next test that Matt set me. This time I had to copy drawings of everyday objects, but after a 15-second delay. There were no prizes for artistic talent – the idea was simply to avoid words and see if I could remember what I had seen. One object was a duck, very poorly copied in my case but acceptable in terms of passing the test, simply because my duck had two legs; a common problem with sufferers of Pick's disease is that their ducks frequently have four. Pick's disease patients recognize that a duck is an animal but, because they are incapable of any further distinction, they overgeneralize and base the drawing on the fact that most animals have four legs. Hence, a four-legged duck.

The evidence from Pick's disease seems to suggest that the temporal cortex specializes in factual memory. If correct, this interpretation would throw new light on some old findings made half a century ago by Canadian neurosurgeon Wilder Penfield. Penfield operated on patients with severe epilepsy, a procedure that entailed exposing the temporal cortex. He was thus able to record the effects of stimulating this part of the brain. In much the same way as Henry stimulated Sarah at the start of chapter one, Penfield applied electrodes to the surface of the brain of conscious patients. Often there was no clear effect, but sometimes stimulation of the temporal cortex caused the patient to experience a memory. In fact, these were not quite like everyday memories – the patients described them as being more like dreams. But perhaps this is not so surprising. An important feature of dreams is that they are not anchored in a precise location or time, so in a sense they resemble factual memories more than episodic memories.

If the temporal cortex deals with factual memory, could another part of the medial temporal lobe be responsible for that other type of long-term memory, episodic memory (memory of events)? Again, recent evidence from patients with more selective brain damage than Harold's indicates that this might be the case, and the precise region involved seems to be the hippocampus, which lies deep below the temporal cortex.

John is 22. During, or shortly after, his birth he suffered a medical condition that starved his brain of vital oxygen for a brief period. For several years he showed no sign of this early trauma, but from age five onwards it gradually became obvious that John could not remember events in his life. And the problem grew worse as he grew older. He now appears to have no autobiographical memory at all, no sense of his own life history. John can recognize himself in a photo taken on his 10th birthday, but he can only do so because he has been told what the picture shows and his factual memory is intact.

So what has happened in John's brain? Scans reveal that his hippocampus is only 40% of the size it should be, so it seems that a short-coming here is to blame. Similarly, three other patients with hippocampal damage have been reported, and each has profound amnesia for events, but each attends a normal school because their language ability and factual knowledge are fine.

John would have had no problem choosing the right type of brush to clean his teeth, but he did have severe difficulties on a test that I too tried at the Institute of Child Health, University of London. Clinical psychologist Claire Salmond showed me a sequence of cards rather like those in the old game of Happy Families. Each card showed a face and name, and Claire told me their occupations. After a brief pause, with the cards put to one side, I then had to answer her questions relating to names and jobs – such as 'what was the clergyman's name?' The idea here was that each time Claire showed me a card, it was a kind of mini-event. The test would assess how well I could recall the details of each event. While I managed reasonably well, John completely stalled.

So it seems that within the medial temporal lobe there is a division of labour: the hippocampus deals with memory of events, the temporal cortex with memory of facts. Perhaps this helps to explain why we cannot trace our earliest memories of life any further back than the age of three. Although some think that this 'infantile amnesia' is due to an inability to use words as memory aids, it must be remembered that most

▲ Marsh tits who stored and retrieved their food in the normal way were found within a few days to have larger hippocampi than those birds who were not allowed to.

of us actually start speaking at around 2 years old. Another idea, therefore, is that the hippocampus is still developing its connections with the cortex while we are very young, so the ability to lay down memories of events takes a few years to develop.

Interestingly enough, animal studies suggest that the hippocampus actually increases in size as a young animal develops the ability to remember events. Oxford University zoologist John Krebs and his colleagues made this discovery after studying the foraging behaviour of fledgling marsh tits. Marsh tits belong to the class of birds that store food in special hiding places and then retrieve it later – a good example of event-based memory at work. In the study, one group of birds was allowed to store and retrieve food in the normal way, and another group was not (they were fed with ground seed that could be eaten but not carried away). After only a few days of storage and retrieval, the birds in the first group had a significantly larger hippocampus. Just as we saw in chapter three, experience can make the brain produce new connections, changing its physical structure in some way to suit an animal's needs.

Other animal studies reveal that the hippocampus does more than just recalling location and time. It turns out that this part of the brain plays an important role in helping animals to navigate by means of a mental map of their surroundings. Rats can normally be trained to learn their way around a maze, provided a reward of food is put in the right place from time to time. But if a rat's hippocampus is removed, it cannot learn its way around at all. Loss of the hippocampus seems to cause two distinct problems: the rat cannot incorporate visual landmarks into its mental map, and it loses the ability to navigate by its own body position – it loses its sense of direction. The neuronal nuts and bolts that underlie these two functions remain vague and controversial. British physiologist John O'Keefe suggests that special cells in the hippocampus might serve as 'place cells', changing their priorities, their pattern of activity, when a rat encounters new landmarks. Another problem is that it is hard to understand how the rat's map sense relates to the human hippocampus. Are humans and rats completely different, or are mental maps a primitive type of the episodic memory that humans possess? Most neuroscientists tend to the latter view. The brains of rats and humans are very similar structurally, so it seems likely they function in a similar way, and in any case, the ability to remember location seems a basic requirement for any animal – be it human, rat or bird – to find its way around.

Even more tantalizing clues about the workings of the hippocampus in memory come from studies of people who suffer from amnesia – the loss of memory. These clues hint that the role of the hippocampus might lie in the laying down of new memories.

There are two distinct kinds of severe amnesia: the inability to remember things that happened *before* the event that caused amnesia (retrograde amnesia), and the inability to remember things that happen *after* the cause of amnesia (anteretrograde amnesia). This distinction has long prompted the idea that separate processes are responsible for laying down memories and subsequently retrieving them. For example, Harold suffered chiefly from anteretrograde amnesia – he became unable to recall events in his life that happened after the operation to remove his medial temporal lobes. This suggests that a part of the medial temporal lobe – perhaps the hippocampus, the area most closely associated with event-based memory – is somehow concerned with laying down new memories. Intriguingly, however, we have also seen that Harold's memory of the two years prior to surgery was impaired. How might we explain this curious phenomenon?

One possible answer is that the hippocampus lays down new memories but holds them only temporarily. After two years or so, the memories are transferred to the cortex for permanent storage. As yet, no one has come up with a nuts-and-bolts explanation of how such a scheme might work. But it seems reasonable to think that factual memories might involve less processing, and so could proceed directly to the cortex without the sustained input of the hippocampus. In contrast, the memory of an episode relies on a host of assumptions, ideas and values that would require far more cross-referencing than, say, the memory of the French word for table. Perhaps a function of the hippocampus is to enter into a dialogue with the parts of cortex responsible for storing facts, so that gradually a coordinated network of all the relevant neurons in the cortex can be brought together. Such a process would take time, and, of course, it might change over time as the associations between facts changed. In this way, the hippocampus might be working like scaffolding on a building – crucial while the edifice is being assembled, but then unnecessary.

But even if this scenario were correct, it would be wrong to think of the hippocampus and temporal cortex as being exclusive centres for memory. Damage to other parts of the brain can also cause sudden amnesia, as a radar technician who we shall call Jack found to his cost some 35 years ago. Jack was the victim of a bizarre accident in which a length of fencing foil was driven right up one of his nostrils deep into his brain. The injury destroyed part of a small cluster of neurons in the centre of the brain known as the thalamus. As with Harold, Jack initially had amnesia that affected the two years prior to his accident, but in Jack's case the impairment faded and the gaps in his memory filled in. Two and a half years on, he has remembered everything up to a fortnight before the accident.

This unusual case of amnesia suggests that the part of the brain that Jack damaged is important for retrieving memories. Studies of Jack's learning skills also suggest that the thalamus might be important for laying memories down. Jack takes longer to memorize facts than most people, but, curiously, he remembers what he has learnt just as well. In contrast, people with damage to the medial temporal lobe, such as Harold, forget much more readily. This discrepancy suggests that the thalamus and the medial temporal lobe play different roles in memory. The medial temporal lobe might be important for actual storage of memories, whereas the thalamus system and an adjacent area, the mamomillary body, are more concerned with laying down and retrieving those memories.

Another area involved in memory is the basal forebrain, a part of the brain that has cropped up a few times already in this book. In chapter one we saw that cell death in the basal forebrain is what causes the dementia and memory loss that characterize Alzheimer's disease. In fact, a range of other conditions can cause dementia by damaging the basal forebrain, including Creutzfeldt-Jacob's disease (the human equivalent of mad cow disease), Down's syndrome, post-alcoholic Korsakoff's syndrome, and repeated blows to the head. In chapter two we came across the basal forebrain again, as the fountainhead of neurons that use acetylcholine, the neurotransmitter that increases arousal. So how does the basal forebrain figure in memory?

It turns out that the basal forebrain's fountainhead has a very pervasive outreach. Some of its neurons run to the thalamus and hippocampus, two parts of the brain that we have just seen play a key role in memory. The basal forebrain supplies its target cells in both these regions with the neurotransmitter acetylcholine, so might the memory loss of dementia, therefore, simply be caused by an interruption in the supply of acetylcholine to the thalamus and hippocampus?

No one knows for sure what the role of the basal forebrain is, but one theory is that it plays an indirect role in memory by increasing arousal. Perhaps acetylcholine simply makes the hippocampus and thalamus more alert, and hence better able to process memory. One only has to think of the effects of smoking to see the attraction of this idea. Ex-smokers are only too aware of the beneficial effects that nicotine, which acts as an imposter for acetylcholine, can have on concentration and mental powers generally. Indeed, it is an arguable fact of the scientific literature that Alzheimer's disease is less common among smokers. Then again, no one would seriously recommend smoking as a preventive medicine – the health risks incurred far outweigh any possible protection from the curse of neurodegeneration.

So far, we have seen that many parts of the brain make a contribution to memory, but as yet we have not touched on what is presumably the very heart of the memory process – the actual storage of memories themselves.

In 1929 the question of where memories are stored was perplexing the pioneering American neuropsychologist Karl Lashley. Lashley found that rats that had been trained to get round a maze had difficulties finding their way after he had made lesions in their cortexes. To find out exactly where the memories of the maze were stored, he made lesions in different parts of the cortex and watched the effects, but to his frustration no differences in behaviour emerged – the rats were equally impaired no matter where the lesions were. Lashley concluded that there was no specialized memory centre and that all parts of the cortex contributed equally to memory. It now seems he was correct in the first conclusion but not in the second. There probably is no particular brain region that acts as a dedicated memory centre, but, as we have seen, different parts of the brain make different contributions to the process of memory.

Let's return to the Canadian neurosurgeon Wilder Penfield, who discovered he could trigger dream-like memories in patients by electrically stimulating the surface of their exposed cortex during surgery. One of Penfield's most intriguing findings was that he could evoke the same memory twice by stimulating different points on the cortex; conversely, stimulating the same site twice sometimes triggered different memories. This remarkable finding shows that memories cannot be tied to individual neurons like the circuits imprinted on a silicon chip; a more plausible interpretation is that a single memory is somehow distributed over a network of neurons, the members of which are free to participate in other networks. When Penfield stimulated the same memory from different sites, his electrode must have been accessing the same network from different vantage points. And when he attempted to stimulate the same site twice, his electrode was tapping into different networks, perhaps because of subtle changes in the chemical environment at that point. This idea of 'distributed memory' can be taken further.

In chapter four we saw how the visual system divvies up the job of processing images – different networks of neurons process colour, form and movement, for instance. This divvying up, or 'parallel processing',

▲ Smoking seems to aid concentration and it has been reported that Alzheimer's disease is less common among smokers.

also takes place in memory, as a simple experiment, carried out by Steven Rose, Professor of Biology at the Open University, reveals. Rose gave chicks coloured beads, some of which were dipped in a bitter-tasting chemical, and found that the chicks very quickly learnt to avoid pecking the bitter ones by learning their colour. Their memory of the bitter beads, Rose found, is stored in two different parts of the brain. When Rose made a lesion in both of these regions at once, the chicks failed to learn to avoid the bitter beads. But when he made a lesion in one of the regions, the chicks learned to avoid the beads but failed to discriminate between colours – they avoided all beads. Finally, when Rose made a lesion in the second region, it had no effect. He concluded that the memory is distributed between at least two parts of the bird brain, one part responsible for colour and the other for size and shape. Although bird brains are quite different from our own, the principles underlying the way they work are probably similar. Perhaps our episodic memories – our memories of events with a particular time and place – are distributed across the cortex in a similar manner to the chicks' memories, the different subdivisions of the cortex processing the many different components of each episodic memory in parallel.

Because we have so much more cortex than other animals, it seems sensible to look here – and especially in our enormous prefrontal cortex – for clues about what gives humans in particular such powerful memories.

It turns out that the prefrontal cortex is involved with a particular type of memory called 'working memory'. This type of memory is responsible for keeping in mind all the information relevant to the task in hand. For example, when playing chess your working memory remembers where all the chess pieces are, and it helps you work out where they will end up as you imagine your ingenious plan of attack unfolding. Evidence that working memory depends on an intact prefrontal cortex comes from an experiment called the delayed-response test. If a monkey is repeatedly offered two beakers, one of which always contains food, it soon learns to reach for the one with food. But if the food is transferred from one beaker to the other while the monkey is watching, and the two beakers are offered to the monkey shortly afterwards, its working memory comes into play and it reaches for the new beaker. However, a monkey with damage to the prefrontal cortex cannot make the necessary mental adaptation and continues to reach for the now-empty beaker.

So is the prefrontal cortex the centre for working memory, and do our prodigious working-memory skills account for the cortex's enormous size? Again, we need to be careful not to ascribe whole functions to particular parts of the brain. For instance, good performance in the delayed-response test also depends on attentiveness (arousal), which, in turn, depends on the basal forebrain – so more than one part of the brain is involved in working memory. Likewise, perhaps more than one function could account for the size of the prefrontal cortex.

One such additional function could be our episodic memory. More than any other animals, humans are particularly good at remembering specific events; it seems unlikely that a rabbit would remember the particular night a fox came by, although the more general memory of its smell or appearance would probably be embedded in the rabbit's memory. In the rabbit's case, the memory of the fox is generic, more factual than episodic. Perhaps our special ability is that we can absorb more memories as events – as well as facts – than can other animals, so we can take in a much greater volume of cross-referenced information. In line with this idea is the interesting observation that damage to the human prefrontal cortex causes an impairment called source amnesia. In source amnesia, memory remains but it can no longer be anchored in a particular time or place – it is no longer an event, but a fact.

A final point to bear in mind is that the boundary between memory and other mental functions is far from clear. Lincoln Holmes, the man who could see normally but could not recognize faces, is a case in point: was his bizarre handicap a disorder of vision or a disorder of memory? It is impossible to separate the two. Although the word memory has a clear-cut meaning in our everyday language, this does not mean to say that it is based on a clear-cut process in the brain.

We have seen in this chapter, and in the two that preceded it, that we know a lot about how the brain becomes individualized. There is no special, single feature; rather, individualization results from interaction between the environment and mechanisms inside the brain, causing changes in the way neurons network together across the whole brain. It is this individualization, which is retained even during sleep and is thus independent of consciousness – that I personally view as the *mind*. As we age and acquire ever more memories, our minds become increasingly personalized and individualized, and better able to understand our world. Perhaps this is what is meant by the wisdom of old age.

That process of individualization is never complete – the brain is ceaselessly dynamic, in constant and exquisitely sensitive dialogue with the outside world. Modelling this sensitivity and dynamism is one of the great challenges facing those who wish to replicate the human mind in a computer.

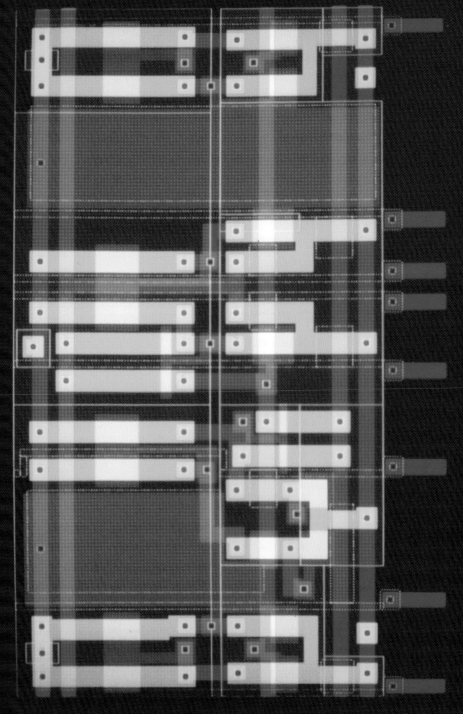

ARTIFICIAL BRAINS

Kevin Warwick and Igor Aleksander have much in common: they are both electrical engineers who build robots. Kevin has spawned insect-like automata that navigate an arena and learn not to bump into objects. Igor, meanwhile, has given birth to Magnus (multi automata general neural units structure). Magnus doesn't move, but when confronted with a printed word – 'cat', for example – it gradually assembles on its display the picture of a winsome feline. Magnus can also look at a picture and identify it as a cat, or at least make the word cat appear on its display. For many, this is exciting, cutting-edge technology; for others, it is the first step towards a disastrous future when robots will take over the world.

A critical factor in a coup against natural intelligence by artificial intelligence would be whether or not the machines were conscious. Non-conscious objects, like stones or beer cans, have no desires – and it is only when you actually *want* to take over the world that you will try to do something about it. So any machine with hawkish aspirations would have to be conscious. Yet here, Kevin and Igor appear to disagree. For Kevin, the whole point of building robots is to make machines that can carry out the unpleasant and dangerous tasks that we conscious creatures don't want to do. Igor's aim, on the other hand, is to build brain-like systems that help us to understand how natural brains work, and to crack the ultimate riddle of how a brain generates subjective experience.

The question of whether non-biological devices could ever be conscious is far from new. The 17th-century French philosopher Julian Offroy de la Mettrie ventured the idea that man is merely a machine; as one of his followers put it, 'the brain secretes thoughts like the liver secretes bile'. Around the same time René Descartes was drawing diagrams of how the brain might work as a mechanical device; he did not include in his reasonings the phenomenon of the soul, which, in keeping with the religious mood of the time, was thought to float free from lowly physical considerations.

▲ Igor Aleksander and the Magnus system. Magnus is a virtual brain model used here to demonstrate how a human might imagine a 'blue banana with red spots' without ever having seen one.
◄◄ A silicon neuron which mimics the electrical activity of real brain cells.

▲ A very early thinking machine, a 'tortoise' that had two responses, attraction to light and avoidance of obstacles.

It was only when programmable electronic computers became available for general use in the late 1940s that pioneer thinkers began to draw serious and detailed comparisons between machines and brains. The neurologist-turned-mathematician Warren McCulloch mused that if one built a computer with as many vacuum tubes as there are neurons in the human brain, one would need the Empire State Building to house it, Niagara Falls to power it, and the Niagara River to cool it. McCulloch was fascinated by the similarity between what was then known about brain cells and the nodes of a computer. He proceeded to build synthetic circuits of 'ideal neurons' – neurons that were unambiguously active or silent, on or off, and which shared a common logic with the operations of a computer.

At about the same time, the mathematician Norbert Weiner coined the term cybernetics. This was an exciting and novel concept, taking its name from the Greek for 'control' or, more precisely, 'steer'. Weiner's idea was to devise principles that could be applied equally to a machine or a brain, and thus reveal how each system worked. In the early days of computers, however, the problem was that the instructions had to be relayed from an external source remote from the main, internal memory. Real brains, though, do not work like that; as we saw in chapter four, our brains are far from being passive recipients of instructions relayed from elsewhere.

But then John von Neumann, a mathematician from Princeton University, devised a scheme whereby the instruction program was actually inside the computer. This advance brought the science of the machine much closer to the science of the brain. Eventually it was even possible to devise a system that could 'learn', not just from simple instruction but from experience, to achieve a desired outcome. The speculation that such devices would one day be as complex, and hence as conscious, as the human brain was suddenly much more believable.

The British mathematician and computer pioneer Alan Turing devised a test intended to be the ultimate way of ascertaining whether a machine was conscious. In the Turing test, a human interrogator asks questions of both the machine and a person without knowing which is which. The goal is to distinguish the two from their answers. The more devious and subtle the question, the more it tends to reveal the difference between some plodding, reflex-like, programmed response, and the agility and ingenuity that characterize the human mind.

As yet, no machine has passed the Turing test. Competitions have been staged in which questions were limited to a fixed range of subject matter, hence making the test much easier, but even so, no machine has managed to dupe the interrogator into thinking that the responses came from a biological brain. Bizarrely, however, the reverse has happened: a human has 'failed' and given the impression that they were actually a machine!

But even if a machine were to pass the Turing test, I, for one, would still object to the conclusion that it was conscious. The Turing test is based purely on responses, yet we know that consciousness can occur without any outward behaviour. Think of someone meditating: the intention is to enhance one's consciousness, yet all outward signs of action are negligible – rather like someone being asleep. Indeed, dreams are themselves a form of consciousness, when the brain is free from the influence

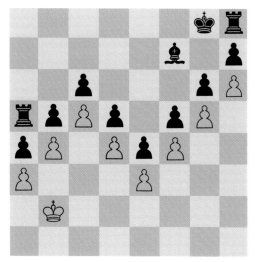

of external stimuli and thus running riot – yet, here again, no outward responses are conveyed to the outside world.

Moreover, computers are increasingly able to enter into convincing, human-like dialogue with people, yet their responses do not indicate consciousness. Personal computers that interact with whoever is tapping the keys, and the perhaps more sinister computer-pets that 'die' if neglected, are powerful examples of how seductive overt responses can be. But no one is claiming these clever gadgets are actually conscious.

The mathematician Roger Penrose has drawn attention to another important difference. Computers function by means of step-by-step instructions known as algorithms. An algorithm is a logical sequence of steps that underpin the strategy for achieving a certain task. For example, many of us learnt at school the sequence of steps, or algorithm, to convert Fahrenheit into Celsius (subtract 32, multiply by 5, then divide by 9). Yet we know that the human brain doesn't work in this number-crunching, mechanical way. Penrose points to the case of the chess computer Deep Thought. Deep Thought was confronted with a chess problem in which it had only pawns left, apart from the king. The pawns were positioned in an impenetrable defensive line, so all a human player had to do to avoid defeat was

Computer scientists in the 1950s became fascinated by the way that networks of living neurons process information by adjusting their connections. To find out more about this process, the scientists set about building electrical networks that worked in a similar way. The result was the neural net – arguably the nearest that computer scientists have got to modelling the brain in silicon.

A neural net is made up of many individual units – the equivalent of neurons – and their connections. In some cases the artificial 'neurons' are electrical components and the connections are wires. However, increasingly neural nets exist not as real physical entities but as virtual systems programmed into computers. Rather like biological nervous systems, neural nets have inputs (the equivalent of senses), outputs (the equivalent of behaviour) and a maze of connections in between. The way a neural net responds to a stimulus depends on the all-important connections. Regular use of a connection strengthens it, making it easier for a signal – a virtual action potential – to pass from one cell to the next. The route that the signal takes through the whole net depends on the pattern of strong or weak connections.

Neural nets differ from conventional computers in an interesting way. Instead of being programmed by a string of instructions, they must be 'taught' to produce a particular outcome in response to a particular stimulus; the teaching session causes the right pattern of connections to become firmly established. Thanks to this ability to learn, neural nets are often used in speech-recognition systems, and they can even be trained to recognize faces or identify the bouquet of a vintage wine – all abilities that were once thought to be the preserve of the human brain.

simply move the king behind the pawns. True, this was not a winning ploy, nor would it make for an exciting game, but the whole point was that the other side could not win. Deep Thought, however, plodding meticulously through the strategies that had been programmed into it, took the rook and thereby lost the game. The computer lacked the common sense to avoid defeat by merely idling away the time indefinitely. Penrose's point is that we use more than the fixed set of rules that are implicit in the problem to be solved. As the physicist Niels Bohr once admonished a student, 'you are not thinking, you are just being logical'.

These non-algorithmic processes, which we call intuition or common sense, are hard to relate to actual physical events in the brain. Yet shouldn't we expect the neuronal maelstrom within our heads to be reducible ultimately to algorithms at the cellular or molecular level? This is the idea of the US psychologist Steven Pinker, who views thinking and feeling as a series of goals and sub-goals that would ultimately be computable if only we could break down these sophisticated processes into the most basic steps. It is merely their complexity, he argues, that gives the illusion that mental processes are not ultimately algorithmic. Let's see if Pinker might be right and explore how easily the basic processes in a biological brain might be modelled in a nonbiological system.

A good place to start is with the brain's building block, the neuron. Physiologists Rodney Douglas and Kevan Martin have built an artificial device that functions in a very similar way to the real thing – they have constructed nothing less than a silicon neuron. When Douglas and Martin stimulate their invention it responds by generating a transient change in voltage just like a real neuron's action potential. If they apply more stimulation, the rate of firing of these artificial action potentials increases – just like in the natural counterpart. Yet, although the silicon neurons are an excellent model for the net activity of a neuron, it is far from being a perfect model of a neuron. As we saw in chapter two, an action potential is really a transient flow and counter-flow of ions. These pass through a cell membrane of extraordinary complexity, not through a semiconductor junction. Moreover, the membrane of a real neuron is just the start of a kaleidoscope of complex chemical reactions seething away inside the cell. These reactions affect not only the influx of ions but also the internal genetic imperatives within the cell nucleus.

Biological neurons thrive on interactions – connections *between* neurons are more important than the neurons themselves. The endless configuring and reconfiguring of connections between neurons is the basis of the brain's nonalgorithmic processing power, and this is the pivotal feature on which brain function depends. So, rather than trying to model the brain with an algorithmic system such as a computer, can scientists model it with some form of artificial, nonalgorithmic network?

In fact, computer scientists have been trying to do precisely that for decades with devices called neural nets. Neural nets depend on electrical connections rather than chemical neurotransmitters to pass signals from one 'cell' to the next. Interestingly, there are some instances in the brain where neurons communicate with each other without the need for chemical neurotransmitters. Neurons can sometimes be fused together, so those action potentials generated in one cell can spread passively to the next, without the need for synapses or chemicals to cross them. This 'electrical transmission' is another means of neuronal communication that is neither so slow nor so energy-expensive as synaptic transmission. In this system, there is no energy lavished on a luxurious range of chemicals, nor is any time wasted while a neurotransmitter is released, shakes hands with a receptor, triggers an interchange of ions in the target cell, and then gets cleared away. Yet electrical transmission is relatively rare in the brain. Why is the energy-expensive and time-consuming chemical method of synaptic transmission so much more widespread?

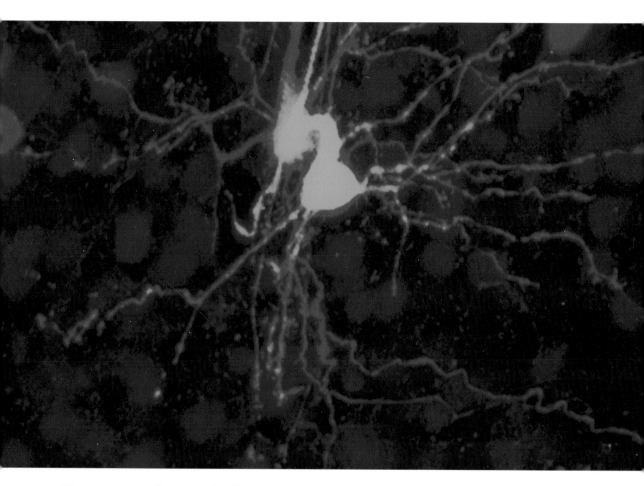

The answer might lie in the flexibility that neurotransmitters give to natural neurons. The artificial 'neurons' in a neural net are simple devices, able to pass on a signal to the next cell or not simply on the basis of the strength of the connection. Real neurons are massively complex by comparison. They have a bewildering range of neurotransmitters at their disposal to pass signals to other cells, and each neurotransmitter can act on several different types of receptor, so its action is not only selective but potentially diverse. The result is an immense repertoire of possibilities that has yet to be reproduced in silicon.

And there is more. So far we have only considered the familiar, well-established process by which one neuron signals to another across a synapse, either exciting or inhibiting the target cell. In recent years neuroscientists have become aware that this is not the full story – there is another dimension to synaptic transmission that is not analogous to the on/off action of computer programs or neural nets. (In fact, it was this

▲ Gap junction in pyramidal cells. A gap junction is the fusion of two neurons forming a low-resistance contact that does not therefore require synaptic transmitters. The current merely flows from one cell to the other. The direct nature of a gap junction can be appreciated when one cell, as here, is injected with a fluorescent dye that then flows into the adjacent cell – a phenomenon known as 'dye coupling'.

simple on/off action that prompted the Nobel Prize-winning physiologist John Eccles to remark in the 1960s that 'everything that goes on in your brain can be expressed in terms of inhibition or excitation'; how easy it would be, given this simple scenario, to draw comparisons between a brain and a computer.) We now know that neurotransmitters can say more than just yes or no – they can 'modulate' a target cell, not giving a signal directly but, more subtly, putting the cell on red alert.

Let's look at an example. When my long-standing colleague Dr Steen Nedergaard from Århus in Denmark applied the neurotransmitter sero-tonin to a cell in the cortex from which he was recording electrical signals, nothing much happened. He then switched to another neuro-transmitter, NMDA, and detected a small change in the excitability of the cell, but not a full-blown action potential. Then Steen tried something else. He gave the cell serotonin followed quickly by NMDA, so that the sero-tonin was still present while the NMDA was applied. This time, action potentials cracked across the screen of Steen's oscilloscope. The serotonin had modulated the cell, making it much more responsive to NMDA.

Over the last decade or so it has become clear that many neurotrans-mitters act not only on their own to inhibit or excite cells, but, given the right circumstances, act as modulators too. These 'right circumstances' depend on the status of the target cell. For example, let's look at an experiment on the effects of acetylcholine on certain cells from the hippocampus. Contrary to what we might expect, acetylcholine has no effect on these cells unless they are *already* firing action potentials. When they are doing so, the acetylcholine removes a kind of molecular hand-brake, and the target neurons keep on firing for longer than they would otherwise have done. So in this example, acetylcholine merely exagger-ates the cells' response. On its own, however, it is about as effective as removing the handbrake in a car that is wheel-clamped.

In modulation, timing is critical – the response of the target cell depends on what has just happened or is happening. Indeed, one way of looking at modulation is as a means of making timeframes important to brain function. Classic synaptic transmission can take place at any time – the neuron's past, present and future are irrelevant. But when we speak of modulation, it is always with reference to two events occurring one after the other within a limited timeframe. And this sophisticated mecha-nism, which doubtless greatly increases the brain's processing power, is yet to be modelled in silicon.

A further discovery that might pose a headache to artificial brain modellers centres on the part of the neuron that releases the neurotransmitter in the first place. The standard story is that neurotransmitter molecules are disgorged from the end of an axon. However, in some brain cells – such as the cells of the substantia nigra, the area damaged in Parkinson's disease – dopamine is released not only in the usual way from the ends of the axons, but also from the dendrites of the receiving cell. This observation, when first reported, was almost unbelievable – after all, dendrites are normally the *targets* for neurotransmitters. In other words, the neurotransmitter is sometimes *going out* through the door marked 'Enter'.

Why do some groups of brain cells operate in this peculiar way? One suggestion is that the backward signal provides feedback to the first cell, regulating its action rather like a thermostat. Another suggestion is that dendrites release neurotransmitter more widely than just into the synapse, resulting in a greater sphere of influence that affects several cells at once. In this scenario, the purpose of dendritic release would not be to transmit a specific message, but would instead have a more general action, such as energizing an entire group of neurons. We have already seen in chapter two that peptide neurotransmitters can be released in this diffuse fashion, so it is not surprising that other neurotransmitters might also behave in unconventional ways. But whatever the explanation, such a promiscuous squirting of chemicals is another effect that makes digital systems look crude and unrealistic by comparison.

Dendrites pose yet more challenges to those trying to model neuron circuitry. For one thing, they are very leaky cables. The successful arrival of a signal at the cell body depends on how long the incoming dendrite is – signals travelling a short distance have less chance to decay and will be more dominant than inputs from further away. Another factor is timing. An input on its own might be too weak to have an effect at the cell body, but two weak signals together may trigger a strong output signal. And, just to make matters worse, neuroscientists have now discovered that dendrites can transmit electric signals without the cell body ever knowing anything about it.

One of the best demonstrations of this phenomenon comes from Professor Rodolfo Llinas in New York. Llinas is particularly ingenious at complex, practical challenges, and here he succeeded in a seemingly impossible task. First he took a thin sliver of brain (less than half a millimetre thick) from the cerebellum of a guinea pig and put the tissue in an oxygenated dish to keep it alive. Next, he deftly inserted an electrode – a thin tube of glass filled with a fluid that conducts electricity – into a single neuron's cell body. In itself, this procedure is rather fiddly, but after

▲ A high resolution fluorescence image, made by Professor Llinas, of a Purkinje cell in the cerebellum of a guinea pig, showing clearly the cell body and dendrites. These Purkinje cells relay signals from the outer layer of the cerebellum to its deep interior.

some practice it becomes quite standard. But then Llinas went on to impale the neuron's microscopic dendrites with four more electrodes, no easy task given that a dendrite is just a few thousandths of a millimetre thick. Thanks to his delicate touch and incredible precision, Llinas made a remarkable discovery: he found that the dendrites were having a party all of their own, generating action potentials independently from the rest of the cell. The signals were not destined for the cell body, they were a strictly local affair.

Later imaging studies confirmed the finding, showing unequivocally that the dendrites of these cerebellum cells can be active independently of the cell body. So what are the mutinous dendrites up to? In chapter one we saw that the cerebellum is involved in learning the kinds of complex movements needed to play the piano or play tennis, for instance. At first, these skills require concentration and input from the senses, but after enough practice they become second nature. It seems that the cells Llinas was studying are involved in this kind of learning. The signals in the dendrites somehow modify the cell so that its response to a certain input – say, the sight of a tennis ball in midair – is permanently altered. More evidence that the cerebellum works this way comes from a classic experiment on learning in monkeys. In this experiment, scientists wanted to see what happened to neurons in the cerebellum when the monkeys learnt a skilled movement. The cells in question, Purkinje cells, receive two main inputs from different parts of the brain. While a monkey is learning a skill, the Purkinje cells must receive signals from both inputs simultaneously in order to fire. But once the task is learnt, only one input is needed. It is as though the joint activity of the two inputs can somehow reprogram the cell so that, in future, only one input is needed. When this is achieved, the response has been 'learnt'.

Of course, artificial neural nets can learn too, but an important distinction is that the mechanism is fundamentally different. Biological neurons don't just change the strength of connections, they grow new connections and let old ones atrophy – they change their physical shape. In other words, hardware is indistinguishable from software.

Some proponents of artificial intelligence (AI) do not worry about the mechanisms involved, and instead put a premium on the learning itself. Just as with the brain, one could argue, Kevin Warwick's mechanical insects and Aleksander's Magnus have the common feature that they become modified by experience. Even a personal computer can learn, although the scale of its dynamism is very different from a brain – computers learn by being programmed in advance with predetermined rules; by contrast, every second is a learning experience for a brain, the internal circuitry of which is continuously changing and updating.

AI proponents would argue that the issue is not to build an exact replica of a brain, but rather to model its essential functions. After all, we don't need to build replicas of birds' feathers in order to achieve flight. In certain ways, AI systems have already achieved this aim. Take, for example, the philosopher-psychologist Paul Churchland, who works in San Diego. He has developed artificial systems that recognize human faces and speech. Similarly, Profesor Rodney Brooks and a team of scientists from MIT (Massachusetts Institute of Technology) have constructed a humanoid robot called Cog. Great progress has been made in programming Cog's arm coordination, vision and his hearing (if he is indeed a male), and it is eventually hoped that his hearing will permit the slow and gradual acquisition of language, a little like in humans.

But the problem, it seems to me, is that both Paul Churchland and Rodney Brooks are putting a premium on a particular mental function –

BRAIN STORY

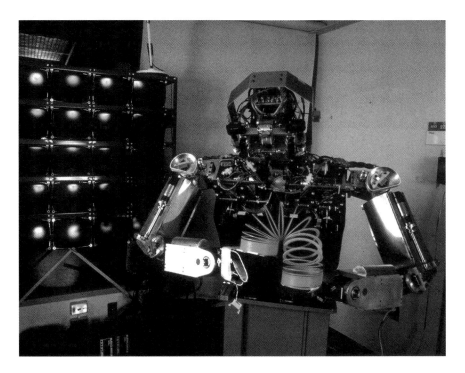

▶ Cog, the humanoid robot which has learnt some basic skills related to hand–eye coordination, such as how to shake its head and to reach for an object.

learning and memory – that computers can *already* achieve more effectively than we can. They are building impressive learning machines without actually providing much insight into real human brain function. The critical issue, the great mystery relating to the human brain, is not that it performs a set of objective tasks – a speak-your-weight machine can do that – but that it feels as well as thinks.

It is going to be really hard to model the human brain in its full biological glory. The artificial version will have to include billions of 'neurons', many with hundreds of thousands of inputs, as well as fountainheads of connections from one major region to another. And the task becomes well-nigh impossible if one also attempts to include, as does Nature, the countless inputs from the rest of the body. Many neuroscientists conveniently forget that the brain is integrated into a body. Think how a feeling (or even a memory) of embarrassment makes your face red, how apprehension makes your palms sweat, how pain can blot out any other thoughts. It is obvious that the dialogue between brain and body is an integral part of the brain's workings. Then there are the longer-term interactions mediated by hormones. These chemicals circulate in the bloodstream and flush through the brain, having all sorts of effects. Hormones are responsible for post-baby blues, for menopausal hot flushes, for road rage, for *crimes passionnels* – not to mention hunger, thirst and lust. To model the brain in its entirety, one needs to model a body too!

But the biggest and most contentious argument about computers concerns their putative inner world, their consciousness. Dan argues that, just because his robot might never have the same level of consciousness as a human, we should not rush to dismiss the idea of machine consciousness altogether. However rudimentary, any form of consciousness would be a promising start. After all, no one would claim an artificial heart is as good as the real thing, but it can do its job.

The problem with this comparison is that the heart has a single and understandable function: to pump blood. However, we cannot easily define the brain's functions, least of all consciousness. Whereas the job of the heart can be reduced to a simple, mechanical bottom line, we are at pains even to understand what consciousness is, let alone model it.

Surely we will not come to understand consciousness by building a machine that we hope will spontaneously transform, like Pinocchio, into something conscious. Yet if we knew what feature to build in, then the problem would already have been solved. And even if we did succeed in building a spontaneously conscious machine, we could not take it apart to see how it worked – we would face the same ethical constraints that apply to humans.

My own view is that the approach of trying to understand the special character of the brain through artificial systems is not very helpful. Understanding learning processes is not the same as understanding that unique property of the mind – feeling. Indeed, even a day-old baby, which has learned far less in its short life than a simple personal computer can learn in a day, has feelings. And so we come to the subject of the next chapter – emotion.

7

THE FEELING BRAIN

'The heart has its reasons that Reason doesn't know', wrote Blaise Pascal. The poet John Keats yearned for a life 'of passion rather than thought'. The Roman poet Virgil described how the Queen of Carthage, on once again seeing her ex-lover, 'felt traces of the old flame creep beneath her limbs'. And the Greek tragedian Euripides, almost 2500 years ago, warned of the importance of the 'wine force' to our mental wellbeing. Humankind has always been in awe of the power of emotion.

The question of emotions is one of the most important that a brain scientist, or indeed anyone, can explore. We are guided and controlled by our emotions. They shape our lives as we attempt to maximize some, such as happiness, and obliterate others, such as fear. Sometimes they seem to take over our whole body and our whole sense of being – we speak of being 'out of our minds'. In a *crime passionnel*, such extreme emotion can sweep away responsibility and judgement. What is happening in our heads at such times? In this chapter we shall attempt to discover what an emotion, in brain terms, actually is.

More than 100 years ago, the English naturalist Charles Darwin noticed something very intriguing about the way we express emotions. In his travels around the world, Darwin saw that people of different cultures convey their emotions with the same facial expressions, even though local languages and customs are enormously diverse. A smile, for instance, means the same thing wherever you travel. Darwin's explanation was that facial expressions are not learnt as a language is – instead, they are 'hard-wired' into our brains, a part of our common evolutionary heritage.

It was not until recently that this remarkable insight was subjected to scientific scrutiny by psychologists, such as Paul Ekman of the University of California, San Francisco. He was amazed that Darwin's groundbreaking work, published in *The Expression of Emotions in Animals and Man*, was ignored for a whole century. In fact, Paul says, it might just as well not have been written.

He filled in some of the background. 'Darwin was a parent – 10 children, I believe, and he watched them, he kept a separate notebook on each child. He saw these expressions burst forth, and he saw them very early,

◀◀ Four of Paul Ekman's photographs of the six basic emotions. Clockwise from top left: fear, disgust, sadness and happiness.

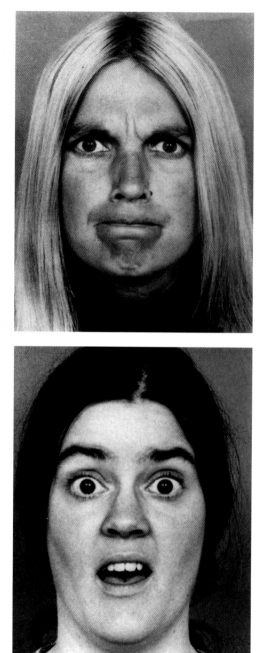

▲ Two more of Paul Ekman's photographs: anger (above) and surprise (below).

and he remarked they were just too early to be the product of learning. Also, doing the voyage of the *Beagle* early in his career, he visited all these exotic places where he didn't share the language, but he felt he had no real problems in understanding their expressions.'

Paul classifies our facial expressions as representing any one of six basic emotions: fear, surprise, anger, happiness, disgust and sadness. He explains how he followed up on Darwin's ideas: 'I really did three kinds of studies. The first wave of studies, I showed faces to people in different cultures: Japan, South America and the United States. I gave them a list of words and, for each face, they had to pick a word that went with it.'

If Darwin was wrong, then the face for anger in Japan might be fear in the US. But that never happened. In fact, in more than 21 different cultures that Paul and his colleagues studied, the vast majority of people gave each face the same name. It looked like Darwin had been right, but there was a problem. Modern cultures share a global diet of film and television that would make it possible for people to learn facial expressions from one another. Paul had to be sure that the responses reflected a common evolutionary heritage, not a common familiarity with Charlie Chaplin or John Wayne, so in the late 1960s he embarked on a mission to find one of the the most isolated tribes on Earth. He found what he was looking for in New Guinea – a tribe so remote from the outside world that the people still lived in the Stone Age. They had no television, no films, no photographs, no magazines, no books; in fact, they were 'pre-literate' – they had no means of reading or writing.

Paul carried out two kinds of study on the stone-age tribe. First he showed them photographs and asked them to point to the ones that fitted particular stories, such as a story in which someone is told their child has died, or another in which someone is about to fight. In each case, the tribe members picked the same faces that we would. Next, he asked them questions such as 'what would your face look like if you were just about to see someone you like?' or 'what would your face look like if someone did something you didn't like?' The expressions of joy, disgust and anger that he saw were exactly the same as in any culture.

His next assignment was in Japan and California. This time he studied facial expressions directly, rather than asking people to identify pictures. He asked his subjects to watch some very unpleasant films while they thought they were alone – but he had a hidden camera. When the scientists examined the precise facial muscle movements later, they found the patterns were exactly the same in each type of expression.

▲ Paul Ekman.

Paul has identified, in all societies, the same facial expressions and the same corresponding six basic emotions – surprise, fear, anger, happiness, disgust, sadness. This common currency of facial expression enables us to express our feelings to other people and to read the emotional expressions of others with immediate accuracy. But we do not just read other people's expressions, we react in a profound and involuntary way. Paul has detected changes in the 'fight or flight' systems of the body – changes in heart rate and blood pressure, sweaty palms, and so on – in people presented with the six basic emotional faces. According to him, these reactions, detected by measuring tiny changes in perspiration in the palms, occur even before the subject has consciously recognized the face. But in some rare cases, this automatic response is missing.

Two years ago, Alan was in a car accident that changed his life. He was not badly hurt, but he sustained an injury to the head that was to distance him from friends and family forever. Alan's wife was also injured in the accident and was taken away on a stretcher. Although she recovered, Alan refused to accept that she had survived. He was suffering from a bizarre and rare disorder called Capgras' syndrome, in which the sufferer believes that the members of their family – or indeed anyone close – have been replaced by sinister imposters.

Experts on Capgras' syndrome interpret Alan's condition as a dysfunction of facial image processing. As we saw with the example of Lincoln Holmes, whose car accident left him unable to recognize faces, there is some kind of system in the brain for recognizing facial images. But there is also a parallel system for processing the emotional content of a facial expression, and this is what was damaged in Alan's case. When Alan's palms were tested for perspiration changes in response to the six basic emotional faces, nothing happened. Remarkably, Alan's subconscious reactions were missing.

For the vast majority of us, this tiny chemical kick occurs whenever we look at our partner or our children. When Alan looks at Christine, he sees a woman he knows looks just like his wife, but without the emotional content the image just doesn't ring true – Alan feels no emotional resonance with her. In an attempt to reconcile this paradox, his brain has come up with the only possible solution: the woman must be an imposter.

In the past, scientists tried to explain our emotions as welling up from a primitive, atavistic part of the brain. Freud regarded the basic drives to destroy and create, encapsulated in his concept of the Id (Latin for 'it'), as the core of mental processes. The demands of the Id were channelled and placed in correct context by the Ego (Latin for 'I'), which, in turn, was suppressed in humans by the morality and conscience of the Super-Ego ('higher I'). In Freud's model, our basic drives were covert, lurking in our subconscious and suppressed by layers of social rules and the expectations of civilization.

When Freud first elaborated his ideas at the beginning of the 20th century, the brain sciences were hardly in their infancy. Freud had started his career as a neurologist, but he quickly abandoned any attempt to root his ideas in the physical brain – too little was known about it, and there were no suitable techniques to test for evidence of the hierarchical structure that Freud was proposing. For this reason, many neuroscientists now regard Freud as non-scientific. On the other hand, he was the first scientist to make a rational attempt at explaining the mysteries of the mind. Freud's description of drives and the subconscious still influences the way we think about the mind, and his technique of psychoanalysis still flourishes, as does much of his terminology in our everyday speech.

By the middle of the 20th century much more was known about the biology of the physical brain. Now scientists could be more confident in applying their ideas to the real three-dimensional structure inside our heads. Accordingly, in the 1950s and 1960s, the US psychologist Paul MacLean developed his theory of the 'triune brain', a theory that had clear parallels with Freud's model. MacLean's main idea was that the human brain could be divided into three easily discernible regions. First came the most primitive part, the brainstem. This is the central part of the brain, where it attaches to the spinal cord. Because the small brains of reptiles have a large and conspicuous brainstem, MacLean interpreted the human version as an underlying 'reptilian brain' that served as the source of our primitive, animal-like urges, rather like Freud's Id. What, then, would be the anatomical basis for the ego?

Wrapping around the brainstem is the limbic system. This was the second stage in MacLean's triune brain, responsible for giving animal-like urges a context. The limbic system is really a conglomeration of interconnected regions of different shapes and sizes, damage to different parts of which can give rise to strange effects on the emotions. Damage to one part of the limbic system (the septum) in rats results in an irrational rage with no apparent cause. Damage to another part (the amygdala) in humans can result in the strange condition Kluver–Bucy syndrome, in which the patient becomes 'hyperoral' and 'hypersexual' – they not only place inappropriate objects constantly in their mouths, but also, more alarmingly, express sexual intentions in an indiscriminate fashion, making advances towards inanimate objects such as items of furniture. Although such behaviour appears highly disturbed, the actual emotions are normal – it is just the context that is wrong.

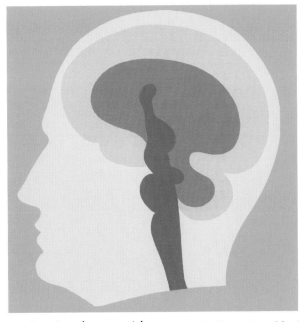

▲ An illustration of Paul MacLean's theory of the triune brain showing the brain divided into three basic regions: the brainstem, the limbic system and the cortex.

The third stage in the triune brain was the cortex. MacLean saw the cortex as the site of logic, conscience and so on, and as such it had the power to suppress the basic urges welling up from the brainstem and limbic system. Like Freud's theory, MacLean's now looks rather dated. Even so, MacLean should be regarded as a pioneer. He did what few neuroscientists have done and what few would dare to do today – he constructed a holistic model of the entire brain.

The idea that emotions are the building blocks of brain function is certainly an intuitively attractive one. We have only to observe a cat purring and kneading a cushion with its paws, or a dog jumping and wagging its tail, to ascribe emotions to animals. Similarly, even very young babies, who seem to have little capacity for complex thought, appear to have a crude portfolio of emotions, such as discomfort and pleasure.

We have already seen that it is misleading to think of brain regions as being discrete centres for specific functions, and one of the problems with MacLean's theory is that it tends to draw us towards that way of thinking. The idea that there are primitive parts of the brain that are normally held in check by some 'higher' centre, but which can sometimes break through, has inspired many interpretations of human behaviour. One example is the mass hysteria of the crowds at Nazi Germany's Nürnberg rallies in the 1930s, interpreted by some as evidence of the reptilian brain unleashed and at work. But the rigid segregation of MacLean's ideas is hard to buy into today. We now know that the supposedly 'higher' cortex is closely linked to emotion – in depression, for example, the prefrontal cortex is unusually active – and the 'primitive' limbic system makes an important contribution to memory.

Another difficulty with MacLean's theory is the assumption, shared with Freud's theory, that emotions are usually suppressed, erupting only in extreme circumstances. Yet our faces are gloriously dynamic – we generate facial expressions most of the time so we must be experiencing some sort of emotion most of the time, albeit to varying degrees. When our faces become blank and unresponsive, then something seems wrong. Indeed, one of the most telling signs of clinical depression is that the patient's face appears rigid and immobile.

Of course, by the time we are adults we are not experiencing the emotional roller coaster of a child, one moment in floods of tears, the next moment gurgling with laughter or shrieking on the cusp of fear and excitement. Yet emotion still permeates our every waking moment – we giggle as we walk along, we whistle cheerfully or sigh with exasperation, we tap our feet with frustration. These tiny signs may be a far cry from a grand passion, yet they testify to some kind of ongoing feeling. After all, we denigrate someone by referring to them as a robot if they do not appear to experience these lesser degrees of feeling.

The constant presence and cultural universality of facial expressions reveals that emotions are a fundamental property of our brains, a basic building block in our mental make-up. Instead of thinking of them as sporadic events that only occasionally erupt, we need to look differently at how they are generated by the brain. How exactly can we relate the constant ebb and flow of emotion, the meshing of feeling and thought, to what goes on in the physical brain?

In the 1950s the American psychologist James Olds carried out an experiment on rats that astonished the scientific community and became a classic. Olds found that if electrodes are implanted into particular regions of their brains, then rats will press a bar to stimulate that region, to the exclusion of all else, even food. Electrical stimulation of the area in question apeared to induce, in animals, some kind of pleasant state since they continued pressing a bar to pass current through an electrode implanted there: a phenomenon now known as self-stimulation. Presumably, the experience the electrical stimulation produced was worth working for and therefore generated some form of rodent 'pleasure'. The area where the electrode was implanted was henceforth dubbed a 'pleasure centre'.

Olds had placed the electrode in a part of the brain called the hypothalamus, a structure like a flat grape that lies almost on the floor of the brain, just in front of the brainstem. Although tiny, the role of the hypothalamus is even more basic than that of processing the senses. Although it receives signals from sense systems such as those for sight, taste and touch, it is also the target for information coming from wide areas of the brain, such the limbic system, and from other parts of the body as well, such as the heart and the gut. The hypothalamus is in a sense bilingual: by means of fast synaptic transmission it is in close communication with other quite sophisticated areas of the brain. But at the same time, the hypothalamus plays a vital role in controlling a most basic aspect of our internal body function: hormones.

Hormones are much slower-acting chemical messengers than neurotransmitters that circulate through the bloodstream, affecting target cells all over the body. As such, hormones are important in body housekeeping and help to regulate hunger, thirst, sex, growth and the immune system, among other things.

So where does pleasure fit into the picture? Another important function of the hypothalamus – perhaps a key to understanding emotion – is that it can trigger arousal.

Arousal is how alert and excited we feel. It is not an emotion itself, but rather a state that often accompanies strong emotion. We experience the sudden onset of arousal when the body's 'fight or flight' system is put into action, in a life-threatening situation for example. The net effects of the alert-system activation are increased heart-rate, decreased digestion, sweating, clearance of airways and other responses that optimize strenuous exercise. All these effects – and others which you are unaware of – put the body on red alert, ready to take sudden evasive or defensive action. They are triggered by the chemicals noradrenaline and adrenaline, which are released in many parts of the body and work both as neurotransmitters and hormones.

But arousal is not just involved in fear, it is a general feature of emotion. After all, whether we are ecstatic with joy or burning with rage, the common factor is the same: excitement. And arousal is not always extreme – there are a spectrum of variations from drowsiness through to intense excitement. Mild levels of arousal may pass unnoticed, but they still cause the tell-tale increase in perspiration on the palms that Paul Ekman was able to detect in response to the six basic facial expressions. Arousal also shows up on an EEG, causing the same frenetic pattern of brain waves that occurs during dreams.

The hypothalamus has been implicated in arousal since observations made some 60 years ago that monkeys with lesions in the lower part of the hypothalamus were lethargic, whereas those with lesions in the frontal part of the hypothalamus developed permanent insomnia. It is now thought that the frontal cells normally inhibit posterior hypothalmic neurons which otherwise contribute to the onset of arousal.

But the hypothalamus is not the only part of the brain that can have these effects. As long ago as 1949, the Italian physiologists Moruzzi and Magoun discovered that stimulating parts of the brainstem can also trigger arousal. Similarly, lesions in the brainstem cause curious defects of arousal. If the brain is separated from the spinal cord by slicing through the brainstem at a particular point, an animal goes into a coma. But if the cut is made slightly further back, the animal develops insomnia. A cut halfway between these two points causes alternating sleep and wakefulness.

When Moruzzi and Magoun first elaborated their ideas about the brainstem, almost nothing was known about neurotransmitters. Now, however, we know that the brainstem is the site of the neurotransmitter fountainheads that permeate higher reaches of the brain. Cuts in the brainstem sever these fountainheads, depriving parts of the brain of certain neurotransmitters.

Let's return to the idea of the pleasure centre. When Olds's rats were pressing the bar to stimulate their 'pleasure centre', they were actually tapping into part of the arousal system via the hypothalamus. So might the same effect be achieved by placing electrodes in their brainstems? Indeed, rats will self-stimulate electrodes here too, and this has enabled scientists to map out all the 'pleasure centres' in the brainstem. Interestingly enough, they all lie along one particular trajectory – the dopamine fountainhead. So dopamine seems to be the key brain chemical in arousal, released not only in the brainstem but also in the hypothalamus. But why should its release also produce feelings of pleasure?

The idea that emotions might be linked to high arousal was first suggested at the turn of the century by US psychologist William James and, independently, by Danish psychologist Carl Lange. The James–Lange hypothesis was, quite simply, that emotions are no more than feedback to the brain from an increased heart-rate, sweaty palms, and other physical effects of arousal. But this explanation is too simplistic. Consider the following experiment conducted by the American psychologist Harriet deWit of Chicago University.

Harriet gave human volunteers the drug amphetamine ('speed'), which increases arousal by boosting the effect of dopamine in the brain. She wanted to know whether the raised dopamine level produced a consistent emotional response, so she told half her volunteers – untruthfully – that they had been given a placebo, an inert substance with no effect. Harriet found that the volunteers who were expecting the drug enjoyed its effects, but those who thought they had taken a placebo simply felt anxious. The same drug, the same arousal, had produced different emotions. So there must be more to emotion than mere arousal.

▼ Prairie voles mating. Oxytocin in the female plays an important role in its bonding with males and babies, and is possibly very significant in human bonding too.

The inability to differentiate between pleasure and anxiety is not the only drawback of the James–Lange hypothesis. For instance, we know that rats are still motivated by fear even when their 'fight or flight' system has been disabled by destruction of the noradrenaline and adrenaline systems. Similarly, people paralysed by spinal injuries that block 'feedback' to the brain still feel emotions. Indeed, the 'feeling' of an emotion such as fear often appears to *precede* the body's arousal responses. A final problem with the James–Lange hypothesis is that humans can feel intense pleasure without being aroused at all, for example when over-indulging in alcohol or experiencing the alleged dream-like euphoria of morphine.

So arousal does not simply equal emotion, though one undeniably plays a part in the other. Something else must be going on. The British psychologist Jeffrey Gray sees the noradrenaline system as being responsible for arousal and the dopamine system as being responsible for reward – though exactly how dopamine relates to reward is not clear. My own view is that dopamine could in some way predispose groups of neurons to network together in a specific type of configuration that causes pleasure, though precisely what this configuration might be and how it might trigger pleasure is open to speculation.

But even if we understood pleasure, it does not automatically follow that we would understand all about the other emotions too. However, by following this line of thought, we come across a different set of clues. We clearly feel different emotions at different times for different reasons. For example, hormones themselves can trigger certain dispositons. Everyone is familiar with the aggression linked to the male hormone testosterone and the irritability of premenstrual tension, for example. But fewer people have heard of the female hormone oxytocin which is now being found to have unexpected actions in the brain and to be directly linked to emotion.

Oxytocin has long been known to be involved in contractions during labour, breast-feeding and orgasm. However, psychologist Rebecca Turner at the University of California, San Francisco, has been intrigued by the recent discovery of oxytocin receptors in the brain. Like dopamine receptors, these seem to occur in areas associated with 'reward'. Rebecca already knew that in the monogamous prairie vole oxytocin plays a role in the initiation of maternal behaviour and the formation of adult pair-bonds. She wondered whether it might also be important in human social behaviour.

She explains how she set about trying to find out. 'Rather than manipulate oxytocin like you can do in animal studies, we decided to start from the other end – to have people experience certain emotional states in a laboratory. We found that when experiencing positive emotion, particularly about relationships, some women experienced surges of oxytocin.'

In her experiment, Rebecca measured blood levels of oxytocin during three different conditions: massage, positive emotion (where subjects were asked to think about an experience of love) and negative emotion (where they were asked to recall a time of abandonment). A general observation was that oxytocin peaked in the pleasurable conditions of massage and positive emotion, but declined during negative emotion. So as well as acting in the body, it seemed that oxytocin was also at work in the brain. But not all the women experienced a surge of oxytocin. Why not?

Rebecca elaborated. 'People are interested in whether or not there are differences between people in terms of how their oxytocin is regulated, and what we found in our study is that women who told us they had more distress in relationships, had more difficulty in controlling their emotion in relationships, tended to have generally lower levels of oxytocin. In other words, they didn't surge as much during levels of positive emotion and they tended to become quite depleted of oxytocin during negative emotion.'

Such a connection between oxytocin and a propensity towards happiness or withdrawal might in the future prompt a different approach to developing drugs for depression. Certainly, it shows that we should not always think in terms of neurotransmitters. One idea is that oxytocin is in a chemical seesaw with the stress hormone cortisol, which could, in turn, be manipulated with drugs. On the other hand, although Rebecca's work raises interesting questions, it is important not to read too much into the data so far. We do not know what other chemicals or hormones might be fluctuating in the different emotional conditions, nor do we know whether the raised level of oxytocin is cause or effect, chicken or egg. And, in any case, because hormones have widespread effects on the body, drugs that modify their action can have unpleasant side effects.

Despite knowing so much about the galvanizing of brain and body that can occur with a flood of hormones or a fountain of dopamine, we have yet to find out what actually happens in the brain as our emotional temperature mounts. A critical consideration is that the emotional reaction will vary enormously according to personal experience.

Psychologist Paul Rozin of the University of Pennsylvania has made studies of disgust which illustrate this point superbly. Rozin offered children of different ages some rather unusual edible objects. Only after a certain age did a child refuse to eat a chocolate doo-doo or drink apple juice from a new and, therefore, perfectly harmless bedpan. Their disgust clearly arose from specific associations learnt as they grew up.

This interplay between learning and raw emotion is an area that Dr Antonio Damasio, Professor of Neurology at the University of Iowa College of Medicine, is currently exploring. He has shown that, even in seemingly logical decisions, covert emotions play an unexpected and important part, and that these emotions are rooted in prior experience. Antonio proved this to me in a test where I was the subject.

Antonio's task seemed simple enough. I merely had to select a card from one of four piles and show it to him. For reasons that were not explained to me, I was either rewarded with toy money or fined as a result of my selection. The idea was simply to maximize my winnings. After half an hour or so, the winning criterion still eluded me. Whatever rule I formulated – using just my right hand, for example, or repeating a certain number of selections from one of the piles – nothing consistent seemed to emerge. My mistake, like that of most of the others, was to assume there was a logical answer.

Antonio had rigged two of the piles to give less spectacular winnings but less heavy losses. Since this was a tendency rather than a hard and fast rule, it was not obvious. Yet he found that, after a while, most people tended to favour the more modest yet safer piles. Why is that? Antonio's theory runs that emotional associations with winning – and, more significantly, the negative feelings of losing even toy money – lurk in the subconscious, producing in the body tiny chemical changes that he calls somatic markers. According to Antonio, we become conscious of these markers in extreme situations – when we win or lose huge amounts, for instance – but most of the time we are unaware of them. We assume we are acting logically, but in fact we are guided by covert emotions. In a sense, these somatic markers are a kind of intuition or gut feeling.

This scenario certainly fits with the idea that emotions are with us all the time, ticking away constantly in the background. Antonio goes on to propose that there is no single chemical for emotion, but that emotion is made up of a whole landscape of chemicals and processes throughout the body that mesh with associations laid down in the brain. In this way, we can see that the old idea of primitive emotions erupting through a veneer of reason can be replaced by a more realistic scenario in which reason and emotion mesh together to different degrees at different times. But we

◀ Susan Greenfield talking with Antonio Damasio about his theories that emotions unconsciously affect the way we learn and reason.

still don't know exactly what those somatic markers actually are, nor what they do in the brain.

Antonio has found that patients with damage to the prefrontal cortex cannot play his card game well. They seem to lack intuition and instead use logic, to their loss. They do poorly at the game, he argues, because the brain is no longer registering the negative subconscious markers related to parting with cash. But that does not mean to say that the prefrontal cortex is the centre for intuition. It is more likely that it plays a role, along with other regions, in orchestrating somatic markers that influence the configuration of global brain states.

American psychiatrist Doug Bremner of Yale University Medical School, who has been studying post-traumatic stress disorder (PTSD) in Vietnam War veterans, has also found an interesting relationship between memory and emotion. Bremner found that many victims of PTSD had memory problems – they had trouble remembering what to buy at the grocery store, what they had for breakfast that morning, and so on. Bremner's theory was that emotional stress had induced damage in the hippocampus (the part of the brain involved in laying down new memories), a phenomenon known to occur in animals.

He explains: 'We've seen there's quite a bit of research in animals showing that stress can cause damage to the neurons of a part of the brain involved in memory called the hippocampus. And this area is involved in learning new information, it also plays an important role in the emotional, storing the emotional significance of the environment we're in – all the cues in the environment that we can attach emotional significance to.'

To test his theory, Bremner took brain scans of the PTSD victims and indeed found a 10–12% reduction in the volume of the hippocampus compared to those of Vietnam veterans who were unscathed. The damage also helped to explain another strange symptom – emotionally loaded 'flashbacks', memories so vivid that Bremner describes them as 'a movie playing in front of your eyes'.

Physiologist Joseph LeDoux, of the Center for Neural Science at New York University, is also interested in how and where emotions are generated in the brain. He is studying the instinctive reaction of fear we experience during a frightening event, such as being confronted by a mugger. Joe's idea is that emotional behaviour is separate from the actual feel of the emotion. The instinctive physical response is, he believes, possible without conscious awareness. This response has an anatomical provenance that is 'quick and dirty' – it does not route via the cortex, but rather via the amygdala.

The amygdala is the area that, when damaged, is linked to indiscriminate sexual behaviour, as we saw earlier in Kluver–Bucy syndrome. By a combination of experiments involving lesions and other techniques, LeDoux has shown that this part of the brain is a pivotal crossroads for coordinating incoming information with outgoing reactions very swiftly. It allows us to act extremely quickly in an emergency, without thinking. Meanwhile, the conscious, subjective *feel* of an emotion is routed via the cortex. There are connections in both directions, between amygdala and cortex, but things are very one-sided: the influence of the amygdala over the cortex is stronger than that of the cortex over the amygdala. Hence, Joe argues, it is easy to let our emotions dominate our thinking, but not quite so easy to do the opposite.

I personally find problems with these ideas, particularly in extending them to a theory of emotion in general. It is hard to see how the same scheme could apply to pleasure, for example. Moreover, as Joe acknowledges, fearful behaviour is not the same as the subjective feeling of fear. Yet the exciting thing about emotions is that they are truly feelings: we *feel* them. Hence the quick and dirty route, even though it may be true, does not help us explain what an emotion is, nor why we feel as we do.

An important feature of emotion, at least human emotion, is that we do not need a loud bang, a roaring tiger, or a thunderclap to feel afraid.

We would be more frightened if someone calmly told us in quiet words that we were going to die at dawn tomorrow. Clearly, then, the cortex must play an important part as well, and in a way that Joe LeDoux has not included.

The search for the parts of the brain that underpin emotion has, to say the least, proved problematic. Although some areas clearly play a role, that role is neither exclusive nor direct. And some key regions seem to involve themselves with emotion while at the same time contributing to completely different tasks. If emotions are with us all the time, ebbing and flowing and meshing with more logical thought processes, then we cannot expect to ascribe them to a confined corner of the brain. Rather, we should try to discover some kind of holistic, net state of the brain that also varies from one moment to the next and correlates to different levels and types of emotion. The trouble is, how would such states be described? Certainly not in terms of standard brain imaging techniques, which merely show us, piecemeal, splodges of transient activity.

The billions of neurons in the human brain are in a constantly dynamic state, networking together to form an endlessly shifting kaleidoscope of different configurations. We know that these changes are wrought by chemicals such as neurotransmitters and hormones, and we know that we can modify these chemicals with drugs. Since psychoactive drugs also have a clear effect on our emotions, could they offer a bridge to link the subjective feel of emotions with global states of the brain? In the next chapter we shall look at both prescribed and proscribed drugs to see whether we can find more clues into how brain chemicals control the way we feel.

DRUGS

Gordon Claridge was in Wonderland. A clinical psychologist at Oxford University, fascinated by troubled minds, he had been eager to try out for himself the mind-bending and, back then in the 1960s, still legal drug LSD (lysergic acid diethylamide). Under the influence, Claridge was stunned to discover that only so long as he sat perfectly still would the world remain reassuringly the same size, with a consistent appearance from one moment to the next. Once he stood up, he had the impression that he was up against the ceiling, very tall, while his assistant had become a midget. Meanwhile, the jazz music he was listening to was becoming painful to hear, and in fact very frightening. His LSD trip was out of control.

Experiences such as these, under the influence of hallucinogenic drugs like LSD, are so intense that they can be recalled decades later. Users remember every detail of how different the world seemed, how inconsistent, and how terrifying. But how does a drug like LSD, or any drug that acts on the brain, manage to distort our normal feelings and thoughts so powerfully? In this chapter, we shall explore what key factors might bring about such a meltdown of our mental make-up.

Whether you snort, inject, smoke or swallow a drug, much of the chemical is likely to end up in your bloodstream. The blood then transports the drug to the brain, where it leaks out of the myriad vessels permeating brain tissue and penetrates the neuron jungle itself. Some drugs are more potent if taken in a particular way. Heroin, for instance, would get broken down by the liver if absorbed through the gut, so addicts inject rather than swallow it. Crack cocaine passes across membranes in the lungs to get to the bloodstream – when addicts smoke, it can rush quickly from lungs to brain so that onset of action is fast. In general, the most addictive drugs are those that act quickest – it is easy to build up an association between the taking of the drug and its action.

▲ Diseased ears of wheat. The drug LSD is taken from the ergot fungus.

◄◄ Pills of one of the most common analgesic drugs, aspirin.

So molecules of a drug, be it a drug of abuse or a prescribed medication, arrive in the brain. From here, much depends on where exactly the substance goes to work. Most drugs that affect the brain target particular neurotransmitter systems, but some are not so specific. Take, for example, caffeine. Caffeine is closely related to the chemical theobromine, literally 'divine leaf', which is found in cacao beans and used to make chocolate. Similarly theophylline, 'divine food', is an active ingredient of tea leaves, as well as caffeine itself. In a sense then, tea, coffee and chocolate each contain active ingredients from the same common family of stimulant substances (methylxanthines), though caffeine is the most potent.

Methylxanthines block an enzyme which in turn causes the accumulation within cells of a chemical that plays an important part, for some transmitter systems, in the opening of ion channels following stimulation of a receptor. One such receptor is that for noradrenaline: caffeine can thus produce a similar response as for activation of a sub-type of receptor for noradrenaline. Since noradrenaline acts not only in the brain but also in the heart, it is easy to see why coffee can increase the heart-rate. Feedback to the brain that the heart-rate has increased is probably an important factor in the feeling of arousal that the beverage produces. But as caffeine affects so many different types of cell, it is impossible to pin down its stimulant action to a single mechanism in the brain. Indeed, it acts directly on other parts of the body too, having potent effects on the bladder as well as the airways.

If caffeine is the world's most widely used stimulant, then the most widely used 'depressant' – an agent that, contrary to the name, is meant to make you feel not depressed but relaxed – is alcohol. Like caffeine, alcohol does not have a very specific mode of action. However, whereas caffeine works inside cells, alcohol actually has a key action on the actual membrane walls of neurons. This 'wall' is far from being a simple barrier that demarcates the cell boundary. Instead, it is a complex structure, a little like a sandwich – two layers separated by a fat-like filling. Because of its structure the walls of neurons enable it to store electric charge as a capacitor, and thereby generate a whole variety of different electrical signals of different duration. Alcohol, and also certain anaesthetics, disrupt this elegant structure so that, quite simply, neurons cannot conduct their electrical signals as effectively and efficiently as normal.

Caffeine and alcohol, then, work all over the brain. But other drugs act on specific neurotransmitter systems. Let's return to LSD, the drug with which Gordon Claridge was experimenting. LSD is a psychedelic agent, a drug that, even in relatively modest doses, gives you hallucinations. A psychedelic agent cures nothing, but to neuroscientists, or anyone interested in how the brain works, it offers fascinating and powerful clues.

Although LSD is the best-known psychedelic agent, and the1960s the best-known decade for exploring such drugs' mind-bending effects, the use of these substances is far from new. Archaeologists have even found in a 4500-year-old grave in South America a tube believed to be used for cohoba snuff, a potent hallucinogenic herb still used by tribes in the same region today. Many drugs cause hallucinations if given in very high doses, but psychedelic drugs exert their powerful effects in small doses that are not directly related to arousal – they neither sedate nor overexcite; yet somehow our perceptions are changed.

LSD was first isolated in 1938 by Albert Hofman, an industrial chemist working in Switzerland. He obtained the compound from a poisonous mould called ergot, which grows on wheat. The aim of Hofman's research was to discover active ingredients in ergot that might have medicinal properties. At the time, the LSD seemed unremarkable and was stored away. But a few years later, in 1943, Hofman was carrying out further tests and some LSD somehow entered his body by accident. Everything in his field of vision wavered and was distorted 'as if in a curved mirror'. Items of furniture assumed threatening forms. He barely recognized his next-door neighbour, who took on the aspect, to Hofman, of a malevolent witch with a coloured mask.

Despite the potent and bizarre effects on his consciousness, Hofman was told not to bother taking the research on LSD any further. But a few decades later, before it was classified as a dangerous drug, LSD was seized upon by psychiatrists and clinical psychologists in the hope that it might lead to a better understanding of the distorted perception that occurs in certain psychiatric illnesses. Could the drug serve the medical profession after all?

Gordon Claridge found that people taking LSD were able to register sensory events, like a flash of light, far more readily than others, yet their subjective reaction to the light was far less emotional. He immediately saw parallels with schizophrenia. However, the LSD model of schizophrenia was not one that would last. In the 1960s – as now – too little was known about schizophrenia to design experiments that revealed anything more than the apparent similarity of LSD's effects to the disease's symptoms. The similarity of an LSD trip to a psychotic state – and LSD's growing popularity as a drug of abuse – soon led to it being banned.

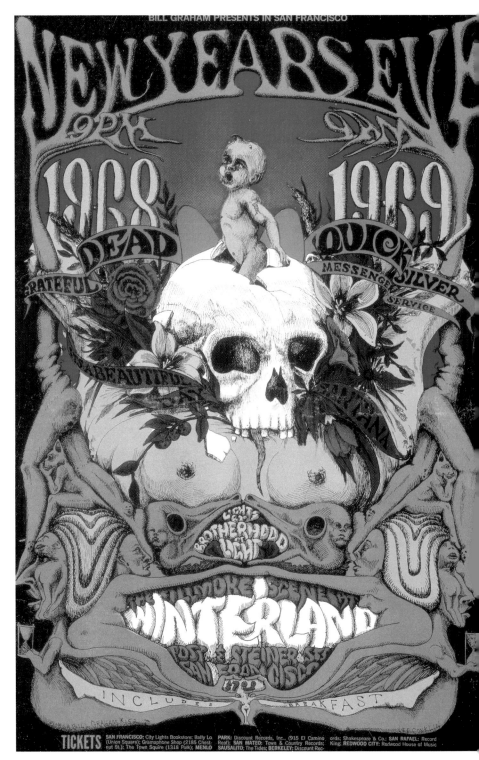

▶ New Year's Eve party poster by Lee Conklin, 1968. Its weird, grotesque imagery is characteristic of the art derived from psychedelic experiences popular in the sixties.

LSD may not have given us any lasting insights into the human condition, but it might still hold clues about the chemical basis of emotion;
after all, when perceptions change, emotions change too. And LSD is
taken illegally because it can induce, in most people, fascinating and
powerful effects. If we track down the site of action of LSD in the brain
we arrive at one of the neurotransmitter fountainheads we met in chapter
two – in particular, the serotonin system. LSD binds to a type of receptor
designed for serotonin and so inhibits the generation of action potentials;
in other words, the drug blocks the effect of serotonin. As it happens, the
drug ecstasy also acts on the serotonin system, but in an opposite way.

Ecstasy first appeared on the streets of Chicago in 1972 and by the
end of the 1980s had made a huge impact in the UK, where it became

▲ Rave culture in 2000.
Teenagers at a club
where the use of drugs
such as ecstasy is
common.

massively popular as a 'dance drug'. It works by triggering an explosive release of serotonin, temporarily flooding the brain with this neurotransmitter. In fact, so much serotonin is released that it overwhelms important physiological control systems, particularly control of temperature and control of water balance by the kidneys. This can lead to dangerous, even fatal, overheating and dehydration. And evidence is coming to light that ecstasy can cause permanent damage to serotonin-using neurons.

Unlike LSD, ecstasy does not readily cause hallucinations, and it seems to induce a sensation stripped of all personalized content. The senses become fixated solely on the throb of the music or the flash of the light, neither of which have any particular meaning. So what can the effects of LSD and ecstasy, and their opposite mode of action on the serotonin system, tell us about how the brain works, how it generates feelings and senses? The answer turns out to be elusive. Other hallucinogenic drugs, such as mescaline, do not act on the serotonin system at all, yet they produce very similar effects to LSD. Perhaps the mind-bending effects of LSD are due to its action on some other, as yet undiscovered neurotransmitter system, or perhaps they are due to inhibition of serotonin somewhere else in the brain. But there may be another explanation. As we've already seen, neurotransmitters work in complex circuits of brain cells that build up into ever more sophisticated brain structures, so their effects depend on the context in the brain; they do not have a function locked inside them. The hallucinations of LSD and the abandonment of ecstasy cannot be explained simply in terms of turning the serotonin level up or down. They are due not to a single neurotransmitter or to a single brain region, but rather to a transient, holistic brain state, some kind of global neuron configuration that can be induced, it appears, by tinkering with the serotonin system.

Long-term users of ecstasy sometimes develop depression, perhaps because of the permanent damage to their serotonin-using cells. So perhaps we can find more clues about mood and emotion by investigating this debilitating condition and the drugs used to treat it.

A core feature of depression is that the patient feels numb – there is an absence of emotion. A massive upheaval must be taking place in the brain to bring about such a devastating flattening of outlook. Is there some single switch that has to be thrown to deaden feelings this way? We do not know the cause of depression but we do know that there are a number of predisposing factors, such as our genes, stress in life, certain personality traits and even diet. But whatever the cause, the medication is the same: antidepressant drugs.

At some time in life one in ten of us will become severely depressed, and in all likelihood will be given antidepressant drugs. The first antidepressants were developed in the 1950s, thanks largely to serendipity. Two independent discoveries pointed to an intriguing idea. The first discovery concerned a new antibiotic, iproniazid, that was being tried out on patients with TB. Surprisingly, the new drug not only helped to treat TB effectively but also caused the side effect of euphoria. The second discovery concerned a completely different group of patients. In this case the problem was hypertension – high blood pressure. Blood pressure can be raised by the neurotransmitter noradrenaline, so doctors gave the hypertensive patients a drug that reduces noradrenaline levels, reserpine. Again, the drug did what it was supposed to do but it had an unexpected side effect – the patients became severely depressed.

What could be the common factor linking these two reports, one of drug-induced euphoria, the other of the opposite extreme, depression? The answer lay in the drugs' effects on neurotransmitters. Just as reserpine depletes supplies of noradrenaline, so iproniazid, the drug that made the TB patients euphoric, enhances availability of noradrenaline. Noradrenaline was the common factor – it was low in the depressed patients and high in the euphoric ones. The obvious conclusion was that noradrenaline, quite simply, made you happy. Not surprisingly, it was not long before this idea gave rise to antidepressants designed to increase noradrenaline levels.

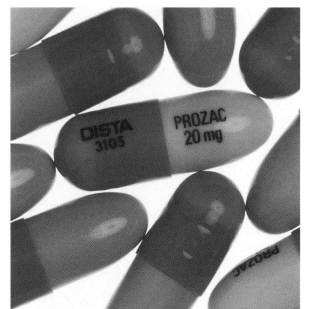

▼ Prozac, the most popular antidepressant drug, acts by boosting the effect of serotonin in the brain.

The first generation of such drugs, called MAOIs, had disturbing side effects and entailed following strict dietary rules. In particular, patients had to avoid mature cheese and yeasty food such as bread. The drugs stopped the body from being able to destroy a toxic compound in these foods, build up of which could lead to throbbing headaches and even brain haemorrhage. Of course, these effects were avoided if the patient followed the dietary rules, but many didn't. Other side effects included insomnia, weight gain and faint spells.

Clearly, the MAOIs were far from ideal. Eventually they were replaced by a second generation of antidepressants, called tricyclics because of their three-

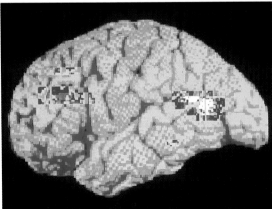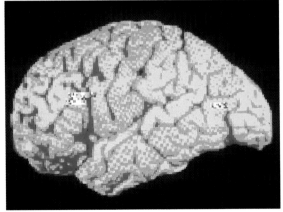

ringed chemical structure. Tricyclic drugs work in a different way from MAOIs but have the same net effect: they interfere with removal of nora-drenaline from the synapse, thus prolonging the neurotransmitter's effect.

But by far the most widely used antidepressants today are a new generation of compounds that target a different neurotransmitter altogether. The most famous example is, of course, Prozac. Like LSD and ecstasy, Prozac acts on the serotonin system, but it works in a different way. When serotonin has crossed the synapse and done its job of stimulating the target cell, it is reabsorbed by the cell that released it and recycled for further use. Prozac slows down this process of reabsorption, allowing serotonin to linger in the synapse and so boosting its effect. Hence, Prozac and similar drugs go by the jaw-breaking name of selective serotonin reuptake inhibitors (SSRIs).

The impact of Prozac on our culture is startling. It is the second most commonly prescribed drug in the USA, whilst in the UK annual prescriptions are hovering around the three million mark. But it is still not clear how Prozac affects mood. We certainly know that, over time, the drug raises the level of serotonin in the brain, and this increase is thought to trigger the therapeutic effect. But then what happens?

Trevor Sharpe, an expert on depression from the Department of Clinical Pharmacology at Oxford University, acknowledges that it would be simplistic to think serotonin was the only chemical involved in depression. On the other hand, serotonin certainly seems to be a key player. One study investigated whether the beneficial effects of Prozac are indeed due to the drug's effect on serotonin level. Scientists gave patients with depression a course of Prozac and allowed the drug to take effect (it takes several weeks for Prozac to work properly). Then they gave the patients a drug that reduces serotonin levels suddenly, even while Prozac is being taken. The result was an immediate relapse in mood. So, if you take away serotonin, Prozac stops working. Hence, the crucial factor is serotonin.

Another interesting connection between serotonin and depression is at the level of diet. Clinical depression does not just affect mood, it can also cause physical symptoms such as insomnia and lack of appetite. As the body requires an essential dietary nutrient (tryptophan) to make serotonin, loss of appetite could compound the problem by depriving the brain of tryptophan and reducing serotonin levels, hence deepening depression. This vicious circle could be a key factor in eating disorders such as anorexia nervosa.

Such has been the enthusiasm for the link between serotonin and

◄◄▲ A polarized light micrograph of serotonin, the levels of which are a key factor in the study of depression.
◄◄ PET scans of the brain of a depressed patient (left) and of a patient who has been-treated successfully for depression (right). The colour coded regions in red and yellow show areas of low activity in the prefrontal cortex (at left) and the parieto-temporal (at right) in the depressed brain. The smaller patches of red and yellow in the treated brain show that metabolic activity and blood flow has returned to the affected areas.

depression that some scientists have extended the idea to explanations of social status. Studies of vervet monkeys have shown that those lowest in the social hierarchy have the lowest serotonin level, while those at the top have the highest level. If the status of a monkey rises, so does its serotonin level. Likewise, in human society the people with the lowest status, such as the poor and women, are more likely to suffer from low serotonin and hence depression, and indeed depression is more common in these groups. One theory runs that depression may have evolved as a means of displaying status to a more dominant member of one's social group – the withdrawn, unassertive behaviour reassures the high-ranker that their potential rival does not represent a threat.

But surely social status and mood are too complex to be explained solely in terms of a single neurotransmitter. We must remember that serotonin on its own does nothing – just as important is its context, the brain circuitry in which it operates and the other neurotransmitters with which it interacts. In any case, there is a danger in explaining behaviour or emotion in simple chemical terms, and that is that the explanation cannot take account of individual responsibility and control. A person may feel there is nothing they can do about it if their brain is deficient for a particular chemical, that they will spend the rest of their life dependent on drugs. Yet statistics show that drugs are not the only factor – many people make a full recovery without the need for drugs, and others do not respond to antidepressants at all.

Of course, antidepressants cannot actually cure depression. Instead, they simply hold biochemical abnormalities in abeyance until a natural remission takes place. A course of treatment allows the chemical environment of the brain to be recast and, in the case of Prozac, entail surprisingly modest side effects. Side effects do exist though. Prozac, for instance, can cause nausea, dizziness and, in men at least, a loss of interest in sex.

Antidepressants are not the only drugs available on prescription to elevate mood. 'Mother's little helpers' were never antidepressants, they were tranquillizers such as Valium or Librium (also known as benzodiazepines), taken to soothe anxiety and stress or to help get to sleep. Instead of targeting serotonin or noradrenaline, these chemicals act on the neurotransmitter GABA, which usually serves to inhibit target cells. Valium and Librium modify GABA's handshake with the receptor in the target cell, boosting its effect and so increasing the inhibition. These minor tranquillizers are the cerebral equivalent of a rainstorm dousing a forest fire.

So successful were the tranquillizers that, for a long time, everyone overlooked the major problem of dependence on these drugs. At the peak of their popularity some 30 million prescriptions a year were being issued for Valium or Librium. The reason so many people continued taking them was not so much because they relieved everyday stress but because stopping taking them caused intolerable withdrawal symptoms. Coming off benzodiazepines can cause intense anxiety, nightmares and restlessness, making life more stressful than ever. Barbiturates were often prescribed as a substitute, but that was like weaning someone off gin with beer.

The modern antidepressants are clearly a better alternative to the promiscuous use of Librium or Valium. But is a drug-induced chemical change in the brain the only answer? For the 30% of people for whom antidepressants don't work, other avenues might get results. One option is cognitive therapy, where a patient and therapist simply talk. The idea is to make the patient see their problems in a different perspective so that they can be more easily absorbed into everyday life and are no longer the cause of an all-consuming depression. Cognitive therapy actually makes serotonin levels rise in some people.

Another approach has been to stir up brain chemistry in an altogether more profound way, not with chemicals but with the sledgehammer of an electric shock. One of the criticisms of this 'electroconvulsive therapy' is that we do not know how it works. On the other hand, ignorance as to how drugs work has never proved an obstacle to their administration. In fact, some effects of ECT have been studied and prove similar to the chemical changes induced by antidepressants. Contrary to the horrors summed up in Sylvia Plath's autobiography, *The Bell Jar*, ECT nowadays takes place only under anaesthesia, but it is still far from being an ideal or pleasant treatment (see also pages 62–3).

What, then, are the prospects for improving on the simple regime of swallowing a tablet of Prozac? At the moment, drugs such as Prozac suffer from two main shortcomings. First, they take several weeks to produce their full effects (which in itself indicates there is more to alleviating depression than just raising serotonin levels). The second problem is simply that antidepressants do not always work – in some people, Prozac triggers a feedback mechanism that causes serotonin-producing neurons to switch off. One aim, therefore, is to develop drugs that block this feedback mechanism, but that would still be trudging along the same old serotonin pathway. Just recently, however, a new avenue has opened up.

As we saw in chapter two, peptides are a totally different class of neurotransmitter. They exist in profusion not just in the brain but throughout the body. Surprising new evidence now links, for the first time, these unconventional and mysterious chemicals with depression. One peptide, substance P, is usually thought of as important in the processing of pain. However, there is now evidence that points to its involvement in stress too. Abnormal levels of substance P might be as closely linked to mood disorders as are abnormal levels of serotonin, maybe more so. So a new type of antidepressant might be one that targets substance P.

Pain has been part and parcel of existence since bodies evolved. Although it serves a useful purpose, helping to keep the body out of danger, it can go on to take over your life. Above all, pain is private, for only you know how it feels.

When part of the body is injured, the damaged skin releases chemicals that activate 'pain receptor' neurons in the skin. Signals buzz along the long axons of these cells to the spinal cord, and then buzz up the spinal cord and enter the brain. It might seem straightforward – you touch something hot or sharp and you feel pain. But even the way the pain signals travel up the spinal cord is far from simple. For example, one route is reserved for the sharp feel of a pinprick, another for the dull ache of stubbing a toe. And even the reasonable assumption that a pain signal will enter the brain is not true.

In the 1960s British physiologist Patrick Wall and his colleague Ronald Melzack developed a revolutionary new theory. Previously, classical theory held that pain was perceived by special receptor neurons distributed around the body, and these simply relayed pain signals to the brain. Melzack and Wall found otherwise. They discovered that the signals from pain receptors had to 'consult' the brain for permission to enter. The brain then replied and either modified or blocked the incoming signal, depending on what else was going on in the brain.

How many times a day does a child bang its knee and come running to a parent asking them to rub it better? This is Melzack and Wall's 'gate theory' in action – the child is literally taking its mind off the pain. Indeed, the extent to which we feel pain can vary enormously. When something more urgent is going on we can ignore pain altogether – soldiers in combat, for instance, can sustain severe injuries without realizing it. And it is not just extreme situations that affect our pain threshold. Even the time of day is a factor – we feel pain least at around noon, for instance.

Sooner or later, though, our pain threshold is overwhelmed and we reach for painkiller drugs. In the UK alone we spend upwards of £300 million a year on over-the-counter pain relief, despite side effects ranging from mild indigestion to stomach bleeding. Go into any chemist and the choice is bewildering. In reality, though, there are only a handful of active ingredients: paracetamol, ibuprofen and aspirin.

Aspirin is the prototype drug in this class of painkiller. It was first marketed in 1898 by Beyer, but has been taken in the form of willow bark for much longer. Despite centuries of use, it is only recently that we have worked out how aspirin actually works. Like paracetamol and ibuprofen, it acts not in the brain but at the site of injury, where it blocks the chemicals that trigger pain receptors. Aspirin does not work for all forms of pain – if you cut yourself it won't stop the wound hurting, for instance. But it is effective at damping down the painful inflammation that develops after an injury. This anti-inflammatory effect also helps to relieve headache by relieving pressure on the brain.

But of all the drugs ever used to combat pain, one substance beats them all and it's been known for millennia. An opium poppy appears on a 6000-year-old Sumerian tablet as the 'joy plant', and in 1500 BC the Egyptians listed opium as one of their 700 medicinal compounds. The use of opium as a recreational drug became widespread in China in the 1600s, creating a voracious market. When the British Empire's East India Company later fanned the flames of this market it triggered the Opium Wars between China and Britain. By 1803 the active ingredient of opium had been extracted and purified to yield a substance 10 times more potent. It was named after the Greek god of dreams, Morpheos.

Morphine is a versatile and powerful painkiller. It can be taken by virtually any route (injected, swallowed or by anal suppository) and in a very wide range of doses. It is readily available, cheap and safe. It does not accumulate in the body, and accidental death with morphine is, in the clinic at least, rare. However, everyone has heard of the addiction that can result from using, if not morphine, then its derivative heroin. Heroin is a chemically modified form of morphine that enters the brain more quickly. Because of its speed of action, it is the addict's preference – it gives them a far faster rush of pleasure; it is more 'heroic'.

If heroin is chemically similar to morphine, does that mean there is a danger that people taking morphine for pain relief will turn into addicts? The short answer is no. As Robert Twycross, an expert on pain management at Oxford's John Radcliffe Hospital, points out, there is a difference between taking morphine for pain and taking it for pleasure. There is a kind of seesaw in the brain where pain and morphine balance each other. If a person is in pain, morphine will restore the balance. But if there is no pain, then morphine – or, more likely, heroin – upsets the seesaw and causes the effects that lead to addiction.

One of the problems with heroin is that its fast action is matched by a fast exit, creating a sudden low that makes the user crave more of the drug. Drugs that wear off quickly, such as amphetamine, often seem to be more addictive for this reason. In contrast, LSD takes up to two days to be eliminated entirely from the brain, and is much less addictive. Hence the rationale behind treating heroin addicts with methadone, a drug that works in the same way but has a more lingering tail off.

In the 1970s scientists discovered that morphine and heroin work by attaching to certain neurotransmitter receptors in the brain. Understanding this process and identifying the receptors was a major breakthrough in brain science. But what were the drugs' natural counter-parts? Evolution doesn't normally equip brains with highly specific receptors for obscure plant chemicals, so it seemed likely that there was another discovery to be made. So it turned out. A short while later scientists discovered the endorphins – natural, morphine-like chemicals that function as neurotransmitters.

Endorphins occur in key parts of the brain involved in pain processing. We know that they normally help to suppress pain because the drug naloxone, which blocks their receptors, increases perception of pain. The 'jogger's high' that some people feel after very strenuous exercise may be due partly to the release of endorphins. But does this mean we could get hooked on endorphins? Not at all. To become addicted to opiates, or indeed to any drug, you need frequent exposure to abnormally high levels of drug. As well as causing addiction, this has another unfortunate effect. Imagine shaking hands with someone very hard or frequently – eventually, your hand will feel numb, and more pressure is needed to feel anything. Similarly, under bombardment from excessive amounts of drug, a receptor becomes less sensitive. More and more drug is needed to have the same original effect, and the outcome is drug tolerance.

A full understanding of the biological basis of addiction remains a key goal for scientific research, but even the definitions of addiction are contentious. Some people, for instance, distinguish between psychological and physiological addiction, while others reject this distinction. The factors that lead to addiction are equally unclear. In some cases one dose can lead to a private hell, while in other cases addiction fails to develop, defying all expectation. The Robins study of 1976 was one of many covering the two-thirds of the US fighting force in Vietnam who were using heroin. This heroin would have been high quality and readily available. Yet when the soldiers returned to the USA only 7% reported addiction problems and attended treatment services. Clearly, the social context of drug-taking is important too.

We set out at the beginning of this chapter to find out whether drugs might reveal the secret of how the brain creates feelings and sensations; but, sadly, we still do not seem to have solved the puzzle. Drugs provide us with enormous insight into how different types of feeling match up with changes in the availability of diverse brain chemicals, but they give little clue as to how such changes are brought about. Perhaps the problem lies in trying to explain sophisticated states of mind exclusively in terms of mere molecules. Of course, transmitters are vital to the rich range of emotions we experience, but they do not have a particular emotion trapped inside them! Instead, brain chemicals, and hence drugs, work in the context of different brain regions, which, in turn, make up the complete brain. The net state of feeling is a product of this holistic brain, involving the coordinated activity of many brain regions. Disembodied transmitters, and the drugs that manipulate them, are necessary but not sufficient for controlling our emotions. The overall working organization of the cohesive brain is the final factor for determining, literally, our state of mind. So, in order to understand the true sophistication of the human brain, we need to search for features that are beyond the nuts-and-bolts level of chemical mechanisms. We need to take a look at the human brain in its entirety.

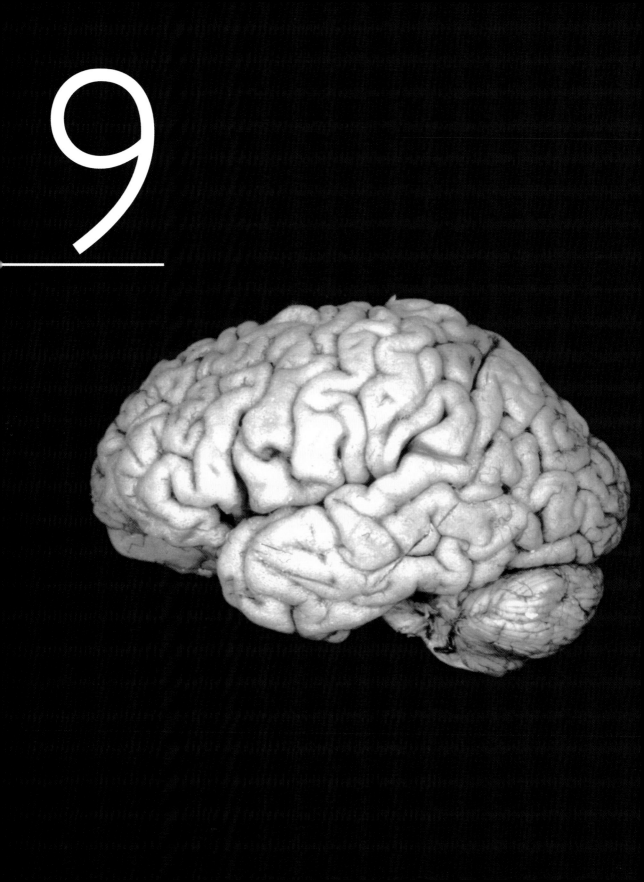

9

FIRST AMONG EQUALS

Tom Harvey handed me a bit of brain about one inch across. It was embedded in hard, transparent plastic like a fly trapped in amber. I found it incredible to think that this innocuous object had revolutionized physics and our understanding of the universe, that it had revealed to us the wonder of space and time. From this particular brain had sprung, indirectly, the nuclear industry and weapons of mass destruction. This brain had made astonishing contributions to our understanding of the immensely large as well as the immensely small. And its greatest achievements came from theoretical reasoning alone, only years later being tested and confirmed by experiment. It was indisputably the brain of a genius. The inert beige lump trapped forever in its plastic casing was, of course, a piece of Einstein's brain.

What was it that made Einstein's brain so special? Where did all that extra intellectual prowess lurk? To discover how to approach such questions – what brain feature to look for – we can ask an even more basic one. Although Einstein clearly had exceptional intellectual powers, the difference between his brain and most people's was nowhere near as great as that between humans and other animals. The most basic issue to explore, then, is what makes the human brain in particular so formidable?

We occupy more ecological niches than any other species on the planet. We have manipulated the natural environment and created our own artificial environments of cities and space stations. Yet the most striking feature about being human is not the external world that we construct but the rich inner world that we inhabit, a world of great idiosyncrasy, replete with hopes and fears, memories and fantasies. Somehow our brains have given us powers of innovation and imagination beyond those of the rest of the animal kingdom. We are undoubtedly a product of evolution, like all forms of life on the planet. But, somewhere along the way, something special happened – and the human brain is the sole beneficiary.

◄◄ This photograph shows the heavily folded cerebral cortex of the human brain.

The most obvious thing about the human brain is its size – it is enormous for a mammal of our body weight. So is size the key to our superior intelligence? In fact, this tempting interpretation turns out to be rather simplistic. Some people, for instance, function perfectly normally without the extra brain volume that seems to be the human birthright.

Daniel Lyon was a porter at Penn station in New York in the early 1900s. When he died suddenly of bronchitis, the autopsy revealed an astonishing fact. Although, in the words of his employers, 'there was nothing defective or peculiar about him, either mentally or physically', his brain weight was almost exactly half the normal adult average, and only slightly larger than that of a chimpanzee.

An even more compelling story is that of Sharon Parker, a 35-year-old British nurse, married with three children. Sharon has led a normal life, but she has a far from normal brain. She tells of how, when she was born, it was immediately clear that something was wrong. Her head had swollen because the cerebrospinal fluid that bathes the brain and spinal cord was not draining away as it should. Her parents were very upset and were advised that it might be better to let her die; such cases of hydrocephaly ('water on the brain') usually result in brain damage. However, they decided to let doctors experiment with a new procedure in which a small tube is inserted in the head, allowing fluid to drain away. So equipped, Sharon went on to lead a normal childhood and, in due course, passed all her school exams. To check that her shunt has remained in place, she undergoes routine scans every few years. These revealed that the central ventricle had ballooned out so much that her cortex was squashed into a thin layer on the inside of her skull – in places only one centimetre thick. Despite this – which, she laughingly reflects, can alarm junior neurologists unfamiliar with her case – Sharon says, 'I would say I'm perfectly normal, I have a pretty busy life. I've compared myself to other people and I can't spot any difference.'

The crucial question is whether Sharon's brain contains fewer neurons or whether, as seems more likely, she has the normal complement of cells squashed together by the fluid pressure in her brain when she was a baby. Imaging techniques allow us to find out. Professor Paul Griffiths at the Royal Hallamshire Hospital, Sheffield, UK, took fMRI scans of Sharon's brain while she performed basic motor and visual tasks. She was also subjected to a battery of neuropsychological tests which confirmed that she had an overall IQ in the normal range, though she had a very specific difficulty recognizing faces. So Sharon's brain is indeed normal, albeit squashed into a tiny volume. Sharon is the exception

▲ Sharon Parker who, remarkably, has led a normal life although she was born with hydrocephaly (water on the brain).

rather than the rule – hydrocephaly usually causes major mental deficits. Perhaps Sharon escaped harmful effects because the condition began late in pregnancy, when the most important parts of her brain had already formed their fundamental connections.

Cases such as those of Daniel Lyon and Sharon Parker raise a puzzling question. Why have we bothered evolving such large brains if we can get by with a lot less? After all, a large head is heavy to carry around, and it means humans must be born at a very early and helpless stage of development in order to squeeze through the mother's birth canal. One idea is that a larger brain gives a greater potential for neurons to form extensive networks of connections. Expert violinists, for example, have a greater cortical area devoted to their left fingers than the rest of us, but they are not born this way – regular practice has stimulated the cortex to form complex new connections.

Certainly, brain size alone is not a sensitive index of ability. The Nobel Prize-winning novelist Anatole France was discovered at post mortem to have a brain about two-thirds the average size, and even Einstein's brain was at the lower end of the size range. Admittedly, there is a correlation between IQ and brain size, but it is only slight, and in any case, many scientists now suspect IQ measurements are a poor test of human intellectual ability. IQ does not measure the ability to understand or appreciate the significance of something, its meaning. This mental capacity to interpret objects, events or people in the light of experience might be equated more with wisdom – the wiser you are, the more meaning you see in life. Surely it is wisdom, not IQ, that makes humans so special.

Wisdom is essential to live successfully in the human world. More than any other animal, we depend on experience – not genes – to give us the skills we need to survive. It is this ability to learn from experience that has enabled people to survive in every corner of the globe, from deserts to poles – one of the unique features of our species. But wisdom is surely not a simple, magic ingredient that appeared suddenly during our evolution. It is more a question of degree. As we evolved and our brains became larger, the ability to form more neuronal connections, and hence to form more associations, gradually increased. Five million years ago, our ancestors were animals much like modern chimpanzees. They led a simple existence, with no agriculture, no computers, no art, no concept of high fashion. It was simply a question of survival. An obvious place to search for the roots of wisdom, therefore, is in the brains of our closest relatives.

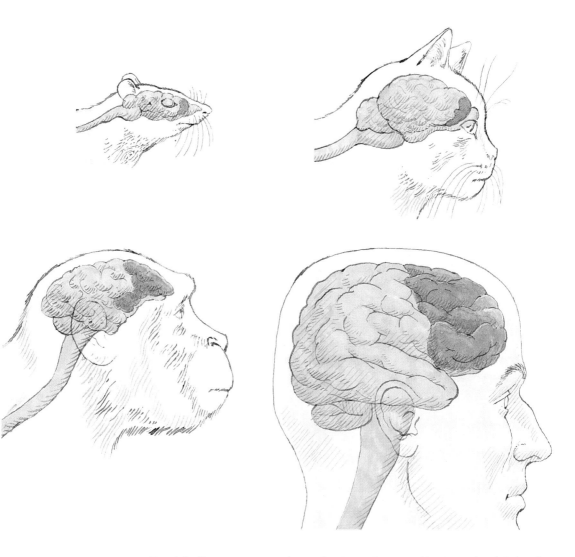

▲ A comparison of the size of the brain's prefrontal cortex (shown in red) in a rat, a cat, a chimpanzee and a man.

Patrick Gannon, an anthropologist at Mount Sinai Medical School in New York, has been studying chimpanzee brains with colleagues at Columbia University. An average adult male's brain is noticeably smaller than a human brain, about a third the volume and with about a third as many neurons. But human brains are not just scaled-up chimp brains – the outer layer, or cortex, is much wrinklier in us. The American physiologist William Calvin worked out the surface area of the cortex in humans and chimps. If flattened out, the cortex of a chimp would be about the size of a sheet of A4 paper. The human cortex is four times bigger (twice what one would expect if ours were simply scaled-up chimp brains). By comparison, a monkey's cortex is only about the size of a postcard, and a rat's would barely cover a postage stamp.

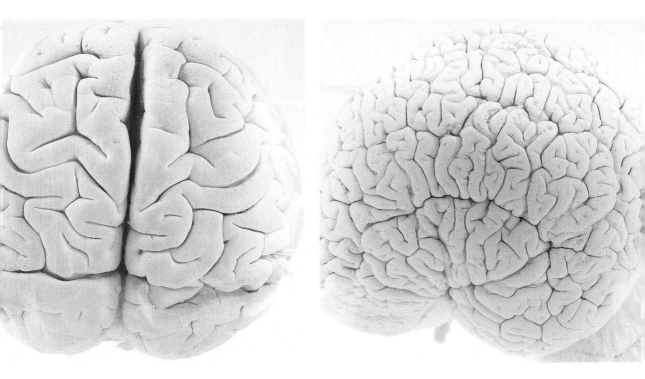

▲ The cerebral cortex of a chimpanzee (above left) and a dolphin (above).

◀ A close-up of a human brain's cerebral cortex. In this photograph, the protective membrane has been removed to reveal the detail of the bulges and grooves that account for the increased surface area in humans.

The cortex is not simply a uniform sheet of cells. Although, to the naked eye, there are no frontier lines crisscrossing its surface, some zones of the cortex have deviated further from the mammalian ground plan than others. In particular, one region has ballooned out of all proportion to body weight during our evolution, growing to twice what it should be for a primate of our size. This is the prefrontal cortex, situated, as its name suggests, at the front of the brain. It accounts for a staggering 29% of brain volume in humans, compared to 17% in chimpanzees and only 3.5% in cats.

The prefrontal cortex has long captured the imagination of neuroscientists because damage to it, in humans, leads to rather subtle problems that have been taken to indicate its role in 'higher' processes. Indeed, on the face of it, the prefrontal cortex is not even an essential part of the brain, but more of a luxury extra, as the now famous case of Phineas Gage illustrates. Phineas was working on a railroad in the middle of the 19th century when an accidental explosion drove an iron bar through the front of his brain. Despite severe damage to the prefrontal cortex, he recovered and was able to conduct himself in a seemingly normal way. The only clear change was in his personality. After a few months it became clear that Phineas was not the man he had been – now he was 'irascible and profane'. No longer able to coexist socially with colleagues, he abandoned the railways in favour of a far more lucrative and easy life as a fairground freak, exhibiting his wound.

This anecdote might have remained just that, but in the late 1930s neurologists developed a treatment that effectively replicated Phineas's accident on the operating table. For the next 40 years or so, up to 35,000 patients in the USA were treated for intractable aggression by leucotomy, a procedure in which certain neuron pathways leading from the prefrontal cortex are severed. The medical profession has, for the most part, abandoned this technique, but the patients remain as testimony to the subtle deficits that arise from frontal lobe damage. As a result of the personality changes seen in lobotomy patients, the prefrontal cortex has been variously ascribed such functions as insight, abstraction, self-awareness and, more gloriously, 'construction and update of representations of the environment'. In short, this region appears to be the pinnacle of our higher processing.

What exactly are the changes in personality and ability that happen when the prefrontal cortex is damaged? A Vietnam war veteran called Michael is, in many ways, a contemporary Phineas Gage. In the late 1960s Michael volunteered for military service while still at high school in the USA. He scored high marks in the qualification test, rose rapidly through the ranks, and was soon winning medals for skills ranging from sharp shooting to good conduct. Then he was posted to Vietnam. One of his jobs in the war was to lead a platoon of 'tunnel rats'. The idea was to flush the enemy out of underground tunnels by lobbing a hand grenade into an entrance and then jumping in to see if anyone was left alive. One night, when out on patrol, he ran into an ambush and an enemy grenade went off just in front of him. The soldier behind him and the soldier in front were killed instantly, yet Michael survived – in fact, he remained conscious. However, a fragment of the grenade had entered Michael's brain in the region of his right frontal lobe. Fortunately, it was only a matter of hours before he was rushed to an operating theatre. Surgeons removed what they could, but some metal stayed permanently lodged in Michael's brain.

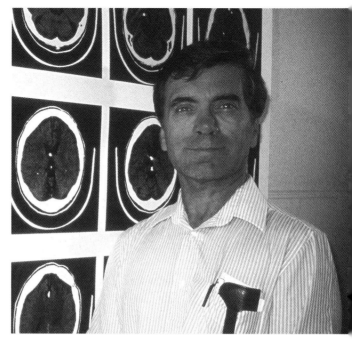

▲ Michael, who survived an injury to his brain during the Vietnam War with his IQ unimpaired, has severe problems with his working memory: he cannot, for instance, weigh up the consequences of his actions or his speech.

After the accident Michael underwent various medical and neurological tests that failed to detect symptoms of brain damage – to all intents and purposes, Michael was normal. But then subtle deficits started to emerge. His neuropsychologist Jordan Grafman, of the National Institute of Neurological Disorders and Stroke, Bethesda, Maryland, explains. 'I think the main cause of Michael's problem in keeping consistent employment has to do with his interpersonal skills. Michael has the tendency to embellish facts, occasionally to confabulate. He's distractible, particularly in social situations. He often makes inappropriate remarks, sometimes sexual remarks that one would not make in private for certain, much less in public. And often he has some difficulty grasping the theme or the moral of a story. If he has to follow instructions, they would have to be extremely concrete. Now having said that, that's remarkable in someone whose intellectual level is not only within normal limits but on many tests well above average.'

Michael's mother tells a similar story. 'He does not have the ability to make plans, follow through with plans, have the judgement to say "we have to pay this bill and we'll have this much money left over". Things

just happen. That's the way his life is, it just happens. No rhyme, no reason, no plans.'

A central issue seems to be Michael's inability, despite a normal IQ, to hold something in mind and weigh up the consequences. In other words, there is a problem with his 'working memory'. Psychologists can detect problems with working memory with a kind of card trick called the Wisconsin card-sorting test. The patient is given cards with colours, shapes and symbols on them, the idea being to sort the cards into piles of the same colour, or piles of the same shape or symbol. After doing this for a while, the rule is changed – the patient might have to stop sorting by colour and instead sort by shape, for instance. People with damage to the prefrontal cortex are unable to adapt to the new rule and continue following the old one.

The psychologist Larry Squire believes that the best way of summing up the contribution of the prefrontal cortex is the 'placement of remembered events in their proper context' – 'working memory'. However, one problem is that even if the prefrontal cortex is important for working memory, it does not have a monopoly over this function. Damage to other parts of the brain, such as the basal forebrain for instance, can also cause problems with working memory.

Similarly, an important dopamine input comes into the prefrontal cortex from a special fountainhead adjacent to the one lost in Parkinson's disease. We have seen that too much dopamine has been linked with schizophrenia; but in schizophrenia the prefrontal cortex shows *under*-performance. What, then, is the excessive dopamine actually doing to neurons in the prefrontal cortex in schizophrenia? One possibility is that

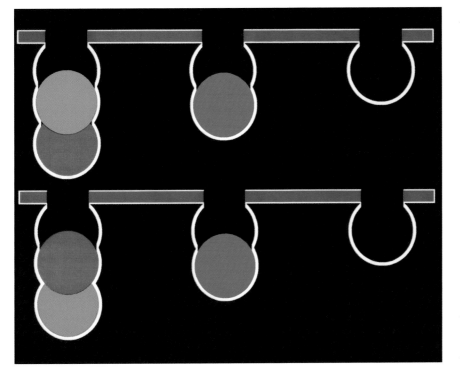

◀ The screen display during Adrian Owen's test, which involves snooker balls, to demonstrate the use of the prefrontal cortex.

it is not acting as an all-or-none transmitter but instead is modulating the response of the cells to other inputs – too much dopamine may cause this facilitating action to stall, as though too much pressure was suddenly put on an accelerator pedal. A healthy amount of dopamine, therefore, is needed for the normal functioning of the prefrontal cortex, but too much ends up having the same effect as direct physical damage to the pre-frontal cortex.

Once again, connections, rather than isolated brain regions, are the watchword. In any case, recent imaging studies have revealed that damage to different parts of the prefrontal cortex can cause different impairments, so to attribute a single function to the whole area would be a mistake. For instance, damage to one part makes a person unable to remember an object's spatial location, while damage to another part makes them repeat the same a task over and over again, even when it has become inappropriate.

Modern imaging technology also gives us the exciting opportunity to watch the healthy prefrontal cortex at work. British psychologist Adrian Owen at the Medical Research Council's Cognition and Brain Science Unit, Cambridge, has developed a test that he believes captures the essence of what the prefrontal cortex is all about. Adrian thinks we use the prefrontal cortex to make mental plans. Imagine going to the super-market, for instance. Rather than taking an actual list, we often go shop-ping with a list of items in mind. To collect them, we plan our route through the shop, mentally ticking off each item as we find it. Obviously, it would be somewhat impractical to take a PET scanner into a super-market, so Adrian has devised an alternative scenario.

Lying in the brain scanner, I listened carefully as he explained what I had to do. Adrian positioned a monitor so that I could reach the touch-sensitive screen. Then he tapped some commands into his computer and a series of coloured balls appeared on my screen. It was as though the balls had been dropped into snooker pockets. The lower half of the screen consisted of three such pockets, each with a different line-up of coloured balls. The upper half of the screen was likewise made up of three pockets containing balls, but in a pattern different from that below. My task was to make the upper set match exactly the sequence of balls underneath. All I had to do was touch the ball to be moved, then touch the pocket I wanted to move it to. Just as with real snooker balls, the top ball in any pocket had to be moved elsewhere before the balls underneath could be repositioned.

I had to do a lot of planning. Sometimes I had to move a particular ball far from where it would end up in order to free up the balls below. Adrian thinks that this indirect, counter-intuitive step is the critical issue. Animals and young children want to see immediate returns – they would place the coloured ball on top into its final pocket, even though such a move would preclude successful completion of the task. Patients with damage to the prefrontal cortex make similar errors. They cannot think through a strategy requiring intermediate steps, especially when those steps seem to be precisely what you do *not* want to do. As a result, they can easily get frustrated by the test – they are aware that their simpler strategy isn't working but they cannot figure out why. For this reason, a time limit is put on the test so that the subject feels, as in my case, that failure is simply due to running out of time.

The PET scanner reveals that the prefrontal cortex is particularly active during this task. When I asked Adrian whether he would go so far as to consider it the centre for problem-solving, he was suitably cautious: 'Now that's a common mistake. What we are looking at here is one part of the network of planning, if you like. So no one part of the brain does one thing and no one part of the brain acts alone. All of our thoughts, emotions and actions are the results of many parts of the brain acting together.' Rather than seeing the prefrontal cortex as the centre for planning, therefore, Adrian sees it as coordinating all the many parts of the brain involved.

The critical question for us here, then, is can other animals form mental plans in this way, or is it an exclusively human ability?

US psychologist Duane Rumbaugh has devoted much of his life to the close study of nonhuman primate behaviour. He has shown that chimps are capable of a certain amount of planning – for instance, they will pick up sticks to use further down the track to dig out ants. Or there is the famous case, often cited as an example of tool use, where chimps strip the leaves off twigs to 'fish' for termites. It seems that our nearest primate relatives can certainly think things through in their heads. The German psychologist Wolfgang Köhler demonstrated as long ago as 1925 that chimps can imagine a solution before acting – he took enchanting photographs of one chimp building a four-storey pile of boxes to reach food hanging overhead.

Duane sums up the situation as he sees it. 'Chimpanzees do plan ahead. I don't believe that they can plan ahead nearly so far as we can. I think also that they reflect upon the past, but not to the degree that we do … I would suggest that chimpanzees are able to plan ahead over the course of several days, whereas we can plan ahead for years or centuries if we wish. I think that that gives us a burden as humans.'

British archaeologist Steven Mithen of Reading University thinks it is one thing to think up a cunning strategy when all the cues and clues are right in front of you – the twig with a few leaves waiting to be torn off, or the boxes left conveniently below a dangling banana – but quite another thing to devise a complex mental plan in the absence of guiding visual cues. Perhaps, therefore, it is the ability to plan in the abstract that sets the human prefrontal cortex apart. In that case, the prefrontal cortex would be not so much a centre for coordination, a brain for the brain, as a centre for *imagination*. It is easy to see how this could have been an evolutionary advantage to our early ancestors. Strategies derived by imagination have enormous survival value – they save time and effort and avoid the danger involved in trial and error.

Another interesting idea prompted by the effects of damage to the prefrontal cortex is the 'source amnesia' mentioned in chapter five. In this case, memory as such is normal, but its relation to location in time and place is impaired. This feature of precise time and space referencing would certainly seem to respect the disproportionate growth of the prefrontal cortex during evolution. After all, although animals can obviously remember, in a generic way, whether an erstwhile neutral stimulus was linked to a rewarding or aversive event, it seems hard to credit them with the ability to pinpoint an isolated, one-off event with the precision that we humans can.

Nonetheless we should be cautious about assigning the prefrontal cortex a unified, global role. We have already seen that different regions seem to yield different impairments when they are selectively damaged, as well as the area that Michael had lost. Conversely, we have seen in chapter four that completely different regions of cortex also play a part in memory, such as the inferotemporal cortex. Damage to the inferotemporal cortex results in a deficit in visual discrimination, even though visual acuity and levels of detection for colour and angles are normal. In particular, memory for familiar objects such as faces is dramatically compromised. Yet, as we saw, the deficit is not one of visual processing itself, but more of memory for pattern differences. The prefrontal cortex does not have the monopoly, then, on any one function, and more than one function can be attributed to it.

One mental skill at which we humans excel is that elusive and tantalizing process, imagination. It may be that some other, as yet unidentified, ability enables us to shut out the press of the moment such that imagination is the consequence, not the actual cause. Steven Mithen thinks the critical issue is that humans evolved the ability to see things in terms of other things, to form associations between objects, people, events. Hence, if a person found a tooth on the ground, they might imagine it as part of a necklace, but a chimp would only ever see it as a tooth. In other words, our ability to form associations allows us to invest *meaning* in things. Meaning, I suggest, is the perception of a person, object or anything else in terms of experience. And the more experience we have, the more meaning we can perceive. Conceivably, this ability to experience a world laden with meaning could be overexercised. Perhaps this is why brain scans of people suffering from depression reveal an overactive prefrontal cortex. Similarly, the blasé attitude of those with a damaged prefrontal cortex might be explained by an inability to attach meaning to events in life.

So does the power of the human prefrontal cortex simply boil down to a greater scope for neuronal connections, allowing associations and layers of meaning so complex that we can look beyond the press of the moment into an imagined inner world? Could more neuron connections be the secret of the human brain?

Let's return to Einstein's brain. Pathologist Tom Harvey and his colleagues at Princeton University Medical Center searched for any distinguishing features they could find in the great man's brain cells.

◀ Pathologist Tom Harvey with Einstein's brain at Princeton University Medical Center. Harvey performed the autopsy on the brain to discover if it had any special distinguishing features.

Only two possible differences emerged. The first was that, in certain parts of Einstein's brain, there was evidence that neurons had been better nurtured than normal. More precisely, Harvey found an unusually high proportion of glial cells in a region that he considered important for sophisticated thought. It was only the ratio that was different, as there were actually fewer neurons than in comparable brains. The problem with this finding is that the brains that Harvey compared to Einstein's were from people who died some 15 years younger. Einstein was 76 when he died. Since we know that the brain can shrink rapidly at this stage in life, it is possible that the change in ratio was simply one of age rather than of genius.

The second observation was that the neurons in Einstein's cortex were packed more densely. Harvey has argued that this allowed the cells to communicate more efficiently, yet studies of rats suggest that neuronal communication is better when dendrites are spaced further apart, perhaps because this prevents cross-contamination and so allows clearer signals. A more prosaic explanation of Einstein's closely packed neurons is that his cortex may have got thinner with age, a natural process in someone as old as Einstein. Although Harvey has pointed out that Einstein showed no sign of the intellectual impairment that usually accompanies a thinning cortex, it is no proof that thinning did not occur.

In 1999 Einstein's brain made the news again when Sandra Witelson, at the Department of Psychiatry and Behavioural Sciences at McMaster University in Canada, claimed that his left and right parietal cortex were both 1 cm (0.4 in) larger than normal. If Einstein's brain was average in size, but this area was larger, then presumably some other area must have been correspondingly smaller. Yet no mention has been made of such a shortfall, nor have any corresponding deficits been found in Einstein's mental portfolio. Witelson also claims that one of the grooves in Einstein's brain (the sylvian fissure) was much gentler than normal, so allowing more connections and better neural communication. Yet, were a smoother cortex to be the key to intellectual agility, surely evolution wouldn't have given humans the most wrinkled cortex of all.

Clearly, these recent claims require more exploration before we can say exactly what made Einstein so special. I personally suspect that the answer lay not in the physical shape and size of his brain, but rather in the processes that occurred while he was alive – the functional configurations that were generated. Since such configurations would have been highly transient phenomena, they would have perished along with their owner and are now lost to our yardstick forever.

As I keep on stressing, we should not concentrate on single cells or single lobes in our quest to understand the brain. We are more likely to find answers by taking a holistic approach. The American biological anthropologist Terrence Deacon argues that it is not the size of the pre-frontal cortex that gives us our unique abilities, nor even the cell connections necessarily, but the balance of power between major regions of the brain. He suggests that our unique intelligence stems from a change in this balance of power that occurred in response to evolutionary pressures on our ancestors. Given the undoubted plasticity of our brains, this suggestion certainly seems to make sense.

Just what was it about our ancestors' lifestyle that made such an enormous difference? Whatever the answer, a clear end-result is that we now have the gift of language. If the root of human intelligence and wisdom is truly to see things in terms of other things, then the invention of words – which can figure in countless different contexts – must have increased our brainpower enormously. Perhaps, then, we have been approaching the problem of human uniqueness the wrong way round. Our brains are not that different from those of other species, and the brain of the greatest genius in history seems no different from any other. Surely, then, it is the processes that occur within the brain that are critical. No process is more human than language, so to language we must now turn.

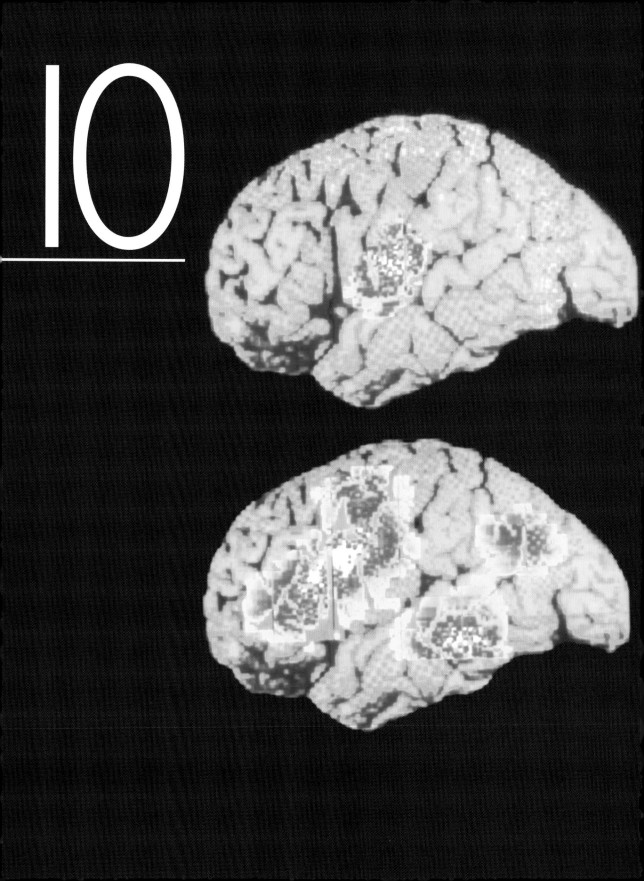

SPEAKING BRAINS

In the summer of 1999 the British archaeologist Tim Flowers made a gruesome discovery – he unearthed a mass burial chamber in South America dating back to the 16th century. He tells of how, horrified yet fascinated, he was confronted with stone walls covered with images depicting a brutal massacre. In hastily dug graves lay hundreds of half-preserved bodies, their contorted limbs frozen in the agony of death…

But now let me tell you that such a place does not exist after all, and, to the best of my knowledge, neither does Tim Flowers. Rather, the point was to demonstrate the power of language – how a few words strung together can conjure up a vivid and riveting scene or the excited face of a archaeologist.

Language is one of the most impressive achievements of the human brain, but does our species have a monopoly on this ability? Many other animals are capable of sophisticated forms of communication. Dolphins, for example, can signal acoustically over miles and can exchange information without disturbing the fish they are hunting. They 'speak' in short phrases, each using an individual whistle rather like a signature. Then there is the complex song of the humpback whale, which can last for up to 30 minutes and carries hundreds of miles underwater. But quite what the songs of whales or the barks, whistles and shrieks of dolphins mean is open to question. Despite years of study, scientists have found no compelling evidence that they make up a sophisticated language in the human sense.

Dolphins and whales use muscles associated with their sinuses and blowholes to make sounds, but land-living mammals use a completely different part of the anatomy. The key organ in land mammals is the larynx, or voice box, a chamber at the top of the windpipe that contains two vibrating flaps of tissue called vocal cords. The vocal cords are normally wide apart, allowing air to flow freely into the lungs and out again. Sound is produced by bringing the cords together and forcing air through the slit in between, making the vocal cords vibrate.

The anatomy of all mammals – apart from humans – makes it impossible for them to speak words as we can. In chimpanzees, for example,

◀◀ Various brain regions linked to language are shown by PET scans which reveal the active areas under certain conditions. Above, areas activated by speaking; below, several areas of high activity generated by thinking about words and speaking.

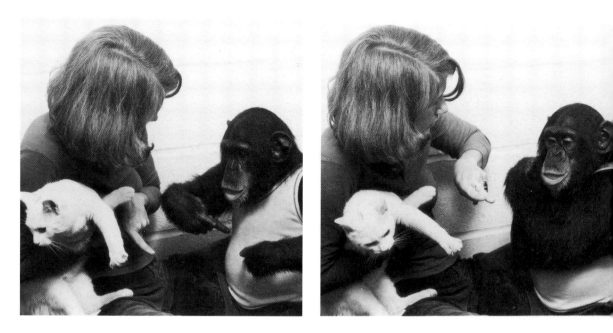

▲ A series of four photographs showing Herb Terrace's chimpanzee 'Nim' who has been taught rudimentary sign language. In the first three pictures he is indicating that he wants to hug the cat. In the last picture, he is given the cat to hold.

the larynx is positioned high up, close to the back of the nose. This efficient design means a chimp can breathe and feed at the same time – when its mouth is full, the larynx rises up like a periscope to connect the lungs to the nose, and food or liquid can pass to either side of the raised larynx to be swallowed. In humans, the larynx is much lower down, making a tight seal between the airways and the route to the stomach impossible. One unfortunate consequence of this arrangement is that food can sometimes fall into the lungs by accident, causing choking. As Charles Darwin wryly observed, humans are 'uniquely adapted to choking to death'. Evolution has decided the risk is worth it.

Human babies have a high larynx like other primates – it is of paramount importance for a baby to be able to suckle and breathe simultaneously. But as they grow older the larynx moves down. The advantage of a low larynx is that we can force sounds through the mouth as well as through the nose. Oral sounds are not only much clearer than nasal sounds, they can also be manipulated by the tongue and the lips.

One of the reasons why attempts to teach chimpanzees to speak have ended in failure is that they are anatomically incapable of producing human sounds. Despite this limitation, there is no reason why chimps should not be able to 'speak' to some limited extent – after all, they make a range of vocalizations to each other. However, there is an important distinction between chimp vocalizations and our own. Whereas we use speech in an intentional and voluntary way, chimp sounds are uttered in

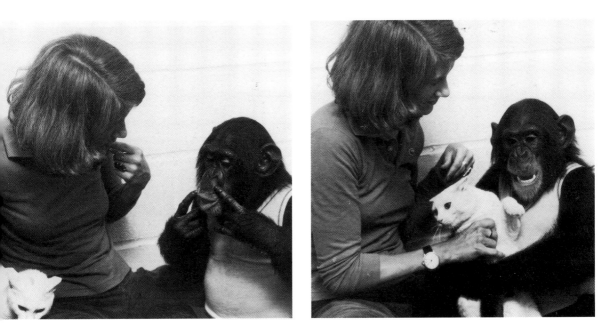

more stereotyped situations, more analogous to the threat and courtship gestures widespread in the animal kingdom. Although it is possible to train a chimp to produce certain vocalizations in new contexts, or to utter them more frequently or less frequently, the production of novel sounds seems beyond them.

Yet language goes way beyond mere speaking – it is a property of the brain. Might chimpanzees, our closest genetic cousins, therefore share with us the ability to think linguistically? Some researchers have tried to get round the limitations of the chimpanzee voice box by teaching chimps sign language or a form of communication based on manipulating symbols. The principle behind such studies is that, although no words are being spoken, there is still the mental requirement to relate a concept or object to an abstract symbol.

Working this way, psychologists seem to have made extraordinary progress with some now superstar chimpanzees. One of the first, Sarah, was able to ask for an apple using plastic symbols; another, 'Nim' (full name Neam Chimpsky, in honour of the great linguist Noam Chomsky) could use American Sign Language to show that he wanted to hold a cat. But fundamental differences between the linguistic abilities of chimps and humans remain. First there is the simple issue of numbers. After four years of intensive training, the chimpanzee vocabulary rarely exceeds some 160 words, whereas that of a human four-year-old is around 3000. Perhaps the most telling difference, however, is that chimps seem only to imitate language in a noncreative way. American psychologist Herb Terrace argues that the critical difference is not so much ability as the *context* in which a chimp uses language. Careful scrutiny of videos of chimps using sign language shows that they do so only to achieve some sort of goal – a hug or an apple, for example. By contrast, even very young children exhibit a spontaneity of speech that does not necessarily mean they want something; we are all familiar with the toddler pointing to the bird, the car, the aeroplane and saying the appropriate word. A final issue to bear in mind is that most of the chimps that use sign language have been intensively trained, although some have picked up the use of symbols just from watching their mothers during training.

US psycholinguist Steven Pinker thinks that language is innate, a part of our human birthright. Whether we end up speaking English, Swahili or Urdu, we all have the same ubiquitous inner language, 'mentalese'. Learning a language, then, is really just a matter of learning how to translate our instinctive, inner tongue into the cultural dialect that we happen to be exposed to. This language instinct evolved, Pinker suggests, through natural selection, just like any other biological adaptation. But there is problem with this theory. As American neuroscientist and anthropologist Terrence Deacon points out, Pinker is merely reposing the question, not answering it. We still do not know how language evolved, nor do we have any precise idea how the brain makes language possible.

But there is no denying that, in humans, language develops naturally and spontaneously. The stages of development are similar in every culture: at 6 months there is babbling; at 1 year there are one-word utterances; by 18 months these single words are being used in context; at 2 years, two-word utterances start to appear; six months later, three-word utterances in many combinations are common; and by 3 years, the child is speaking in full sentences. Finally, by 4 years, the young human has a linguistic ability close to that of an adult.

If language is indeed a biological adaptation, what were the evolutionary pressures that made it develop in our ape-like ancestors? To answer this question, we need to cast our minds back several million years to the prehistoric savannas of Africa.

Scientists currently believe that our tree-dwelling ancestors took up a life on the plains in response to a change in the Earth's climate. As the climate dried, the humid tropical forests that once covered much of Africa went into decline, forcing primates to adapt to new niches. By around 3.5 million years ago, our ancestors were living in the savannas and walking on two legs – perhaps an adaptation to reduce the body's

◄ A rhesus monkey grooming its young: a primitive form of social communication.

exposure to the hot sun. Yet despite their human posture, these upright apes, called *Australopithecus*, were very much like chimpanzees, with a similar brain size and similar diet.

By 2.5 million years ago another type of upright ape had appeared – *Homo erectus*. Its physique was now very similar to a modern human's, suggesting an athletic lifestyle and hairless skin, but its brain was much smaller – about halfway between a chimp's and our own. As ample fossil evidence indicates, this animal put its hands, now freed from the ground, to good use. *Erectus* was a skilled tool-maker, able to make stone 'hand-axes' by chipping flakes off carefully chosen pieces of rock. And once our ancestors had learnt to make stone tools – particularly sharp ones – life would have changed dramatically. Armed with their hand-axes, they found it easier to kill and butcher large prey, and meat became a more important part of the diet. The teamwork involved in hunting, and the sharing of the large carcass afterwards, predisposed these ape-people and their descendants to live in increasingly large social groups.

This is where language comes into the story. Nonhuman primates maintain social bonds within their groups by one-to-one grooming sessions, but this becomes increasingly time-consuming in large groups, and bonds become difficult to maintain. Language may have evolved as a kind of oral grooming, allowing groups to grow larger. Oral communication can reach several people at once, or even everyone at once, as it does in the case of the alarm calls uttered by monkeys to warn each other of birds of prey and snakes (each of which, incidentally, elicits a completely different call).

Palaeontologists can investigate when we started to speak in earnest by exploring when the larynx descended. Although any trace of flesh has long since disappeared from fossil skulls, it is possible to work out the position of the larynx from the angle of the jaw bones. Using this index, it seems that, since the time of *Homo erectus*, our skulls and the brains inside them have been on the move. The larynx began to sink to a lower position, and the brain and spinal cord developed the structures needed for the sophisticated control of breathing required in speech. But *Homo erectus* did not have these features. In spite of having a modern human physique and stone tools, it was still to evolve language. The current thinking, therefore, is that true language did not begin until some time within the last half a million years, after *erectus* disappeared and the more socially sophisticated *Homo sapiens* took its place.

An important consequence of living in a large group is that it can endanger the interests of your genes. Terrence Deacon has advanced the idea that if you were a male who went out hunting for the whole group, you would have a problem. How would you know that a pregnant female that you had mated with was carrying your genes? As a result, pair-bonding would have developed to ensure that you had exclusive mating rights to a particular female. In turn, pair-bonding would have had an important consequence – the need to devise specific symbols to demonstrate overtly an exclusive pair-bond relationship.

It is this ability to use symbols, and, in turn, to treat words as symbols for things that are not present in the immediate vicinity, that appears to be uniquely human. While other primates and very young children can use their general intelligence to communicate in what has been dubbed a 'proto-language', humans alone, after the age of 2 years, seem capable of structuring complex sentences unaided by visual clues. American physiologist William Calvin suggests that this structuring allows us to understand and devise complex relationships between unconnected concepts. First, we see things as symbols for other things – for example a necklace could be used as a symbol that a certain woman belonged to a certain man. Second, we start to think metaphorically, to see one thing in terms of something else, as in Steven Mithen's example of the tooth – a tooth is just a tooth to a chimpanzee, but to an early human it could be a bead in a necklace. Mithen takes the argument even further. He suggests that the explosion of art and culture that occurred about 40,000 years ago can be traced to the dawning of this new ability, not so much to speak as to perceive things *in terms of* other things.

Above all, of course, language has enabled us to exchange and share information, to pass on experience from one generation to the next, so that each new generation does not have to start from scratch. Thanks to the accumulated knowledge of our forebears, we have devised remarkable technologies, built incredible artefacts, and created art, music and literature – all of which have widened the gulf between ourselves and our ape cousins. And that process continues today at a mind-boggling pace. Finally, and perhaps less obviously, language liberates us from dependency on the here and now – we can escape into fantasy, indulge in nostalgia, even contemplate our own deaths.

Surely such a powerful new function must require a different type of brain, or at least some special new feature? The first evidence for a centre for language was reported in the 19th century by French physician Paul Broca. One of Broca's patients had a curious language defect – although he could understand everything that was said to him, the only thing he could say back was 'tan'. Hence, Tan became his nickname. Tan died

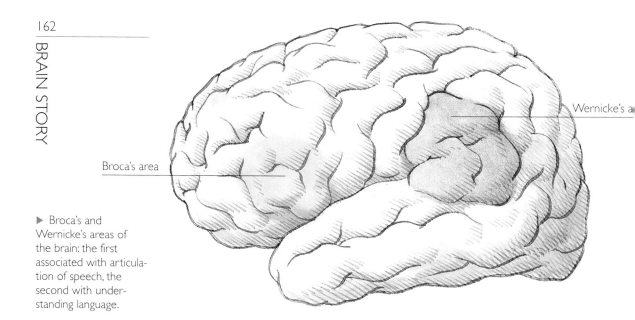

Wernicke's a

Broca's area

▶ Broca's and Wernicke's areas of the brain: the first associated with articulation of speech, the second with understanding language.

▼ PET scans showing a healthy brain (left) and that of a stroke patient. Areas of high brain activity are shown as red and yellow, low activity as blue. The arrow indicates a lesion, an area of brain damage, which has led to partial speech loss.

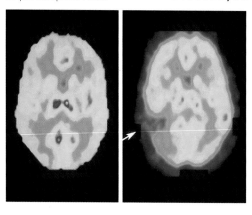

almost as soon as Broca had documented his case, which gave Broca the opportunity to investigate his brain for signs of abnormality. To his intense excitement, he found a large hole in the left side of the brain in an area that now bears his name, Broca's area.

But Broca's area is not the only region where damage impairs speech, as the case of Dr Wilson Talley indicates. Dr Talley, an award-winning American physicist, suffered a stroke in March 1998. His wife Helen sums up on his behalf what happened as a consequence. 'Before you had your stroke you were able to say anything you wanted to, and you could talk, you could teach, you could go to a meeting, and you could do anything and you never thought about it. Then you had the stroke and all of a sudden you couldn't say the right word. You knew the words up here, but when you tried to say it, it wouldn't come our right.'

What does such a language impairment actually sound like? Dr Talley recalls his fateful stroke. 'I has a stroke and crawled to the bathroom, to the bathroom, no to the baseroom and crawl, crawl, crawl and I'm not sure if maybe 5, 10 or 15 minutes, I'm not sure and cold, very cold. It came about 6 hours and tried to shower [pointing to face] this one and only this one perfectly. People, the students were here. 10, 10, 40.'

When Dr Talley tries to talk the words come out fluently – it is just that what he says does not have a lot of content to it. He has difficulty finding the words that he wants, and he has problems producing sentences that are coherent and reflect what he wants to say. But the problem is not just one of expression, but comprehension too.

Helen Talley again: 'You can't just throw out a question as you normally would – it has to be a word-by-word process. You take for granted asking someone what they want to drink with dinner, but you have to completely rephrase and say "Would you like a glass of water?" And then you have to wait for that to click in, as to what water is. Or you might have to just say "Let's go and look at the different beverages" so it becomes a more visual communication.'

American clinical psychologist Nina Dronkers has been working with Dr Talley and explains that his speech problem, or 'aphasia', is different from Tan's. 'What Dr T. has lost', explains Nina, 'is that tool that allows him to go on to the more complex, abstract level of thought. He does very well in thinking out many problems without the use of language – that's very clear. He does maths problems very well, he does certain physics problems very easily. But he does get stuck at a point. It's clear that, without language skills, those more difficult physics problems and those more difficult maths problems still elude him. Of course, it's our hope that as he regains his language skills we'll also see an improvement in his ability to reason.'

Dr Talley's speech impairment is called Wernicke's aphasia after the 19th-century German physician Carl Wernicke, who first described it. Unlike Broca's aphasia, people with Wernicke's aphasia can articulate perfectly well but the content of their speech is garbled. The corresponding part of the brain, Wernicke's area, is also on the left, near the temples.

However, Nina Dronkers cautions against concluding that language ability is controlled exclusively by these two areas. 'Since the time of Broca and Wernicke we've actually learned a fair amount about the language system and the areas that subserve those language functions. We've learned that, yes, Broca's and Wernicke's areas are involved in some aspect of the speech and language process, but the picture is actually much more complex and interesting than that. We've learned that there are several other brain areas that also contribute to different parts of the language process, and we've learned that these interact with areas that help us to do things like focus attention on the task or draw upon certain memories.'

▲ Kanzi, a bonobo chimpanzee, with Sue Savage-Rumbaugh, the principal investigator of the chimpanzee language project at the Language Research Center, Georgia State University. She taught Kanzi to communicate with humans using a computer keyboard. By pointing at symbols that represent various words, Kanzi can construct simple requests just like a young child.

Such findings come from recent brain-imaging studies that reveal the healthy cortex in action while the brain processes language. Thanks to such studies, scientists can watch the cortex at work while a person speaks. One of the most startling discoveries from such research is that saying just a single word causes a unique pattern of activity to ripple across the cortex. The experience of the word 'screwdriver', for example, causes a part of the brain called the motor cortex to light up. The motor cortex is involved in controlling movement, so perhaps this word triggers memories of handling a screwdriver to become active. Obviously, language cannot be the preserve of just Broca's and Wernicke's areas – it involves an eruption of associations and memories that are different for every word.

These scanning techniques now permit us, at last, to explore directly whether chimpanzees process words in the same way as we do. Psychologist Duane Rumbaugh at Georgia State University Language Research Center carried out a study on a chimpanzee called Panzee, who had been trained to communicate by touching symbols. Rumbaugh found that the parts of Panzee's cortex that light up during a language task are very different to those that light up in a human, which suggests that chimps process words in a fundamentally different way. The chimp's brain showed no evidence of asymmetry, unlike humans.

Rumbaugh's colleague Clint Kilts at Emory University tries to explain the difference. 'I think these studies point to the idea that, while the chimpanzee is neurally and perhaps behaviourally capable of language, it does not organize that behaviour in a human-like fashion. From an evolutionary perspective, this may mean that the process of natural selection that favoured the innate ability of humans to use language has not been realized in the chimpanzee.'

So what's going on? As Clint explains, higher primates such as chimps have the same kind of brain plasticity as humans, but their plasticity is general, not specifically adapted to language. In other words, Panzee is relying on her general learning abilities to acquire a kind of simple language, rather than coming equipped with specialized neuronal circuitry as we do.

Another big difference between us and other primates – and one that may be a clue to how and why we are a linguistic species – is that our brains are consistently lopsided – hand someone an object and they will probably take it with their right hand. While other primates may be right- or left-handed, we are the only species with a consistent bias to the same side. Accepted wisdom has it that language is the preserve of the left side, or hemisphere, of the brain, the side that is damaged in both Wernicke's and Broca's aphasias. The left hemisphere also tends to be larger than the right, and this asymmetry has even been seen in *Homo erectus* skulls a million years old.

Mike Gazzaniga, Professor of Cognitive Science at Dartmouth College, USA, is a champion of the left hemisphere. He believes that all the interesting and unique talents that humans have stem from this side of the brain. He introduced me to his patient Joe.

Joe was severely epileptic as a result of an injury, and he had so many fits that he simply couldn't function. To put a stop to the fits, surgeons took the drastic step of severing the connections between the left and right hemispheres of his brain. (This stops the spread of abnormal electrical activity that causes a seizure.) The two halves of Joe's brain are now free to function separately. Although Joe is cured of epilepsy and is to all intents and purposes completely normal, it is possible to detect subtle deficits caused by the breakdown in communication between the halves of his brain. For instance, if Mike holds a picture of a knight in the left side of Joe's field of view (so that the image is processed by the right side of the brain) Joe might say something like 'I have a picture in mind but can't say it. Two fighters in a ring. Ancient and wearing uniforms and helmets … on horses … trying to knock each other off. Knights?'

Mike's conclusion is that language is rooted essentially in the left hemisphere, leaving the right the modest ability to process at best simple two-word sentences. He thinks the left hemisphere makes up stories to explain the actions directed by the right side. As Joe himself described, 'it is hard to explain what it feels like – I do really feel like I'm guessing'.

Because the left side is no longer in communication with the right, it is denied the real inputs that are presented to the right and therefore has to make the best of a bad job by coming up with some plausible explanation. For example, Joe was shown the words 'hour glass' and asked to draw this object. The left hemisphere saw only the word 'glass', yet Joe drew a clearly recognizable hourglass, complete with grains of sand. Why? His answer was that he and Mike had previously been talking about clocks and time. A plausible explanation, perhaps – but one covertly prompted by his right hemisphere.

But this confabulation may be more than just a medical curio. In normal brains, such 'guessing' might be essential for handling the vast amounts of information that rushes at us as we interact with the outside world. It provides a kind of mental shortcut, a crude means of avoiding time-consuming and pedantic data-crunching. Indeed, there is evidence that humans rely more on mental shortcuts than other animals – mice, for instance, seem less susceptible to optical illusions than ourselves. Mike suggests that the human brain has paid a price for evolving language: something has had to be sacrificed for language to fit in.

Mike sees the left–right distinction as going far beyond language. As he explains, in split-brain patients the left hemisphere's problem-solving ability doesn't change a

▼ Watching Joe, whose two brain hemispheres have been separated, doing a 'split-brain' test. Each hand can independently draw a different diagram.

whit, yet the right hemisphere, as he puts it, 'is sort of dumb'. In Mike's opinion, this difference suggests there are specialized circuits in the left hemisphere for cognitive capacity, problem-solving and intelligence.

I had many questions for Mike that took us to the frontiers of research into the fascinating subject of brain lateralization. First we discussed the mysterious 'anterior commissure', a minor pathway between the front of the two hemispheres that stays intact in split-brain patients. Does this secondary pathway take over the job of transferring information from one side to the other? In monkeys, apparently yes. Split-brain monkeys turn out not to be split-brain at all until their anterior commissure is severed. In humans, however, deficits occur when the pathway is intact. So what does it do? As yet, the question remains unanswered, but perhaps it would be rash to assume the hemispheres are 100% independent in human split-brain cases.

Another issue we discussed was the plasticity of the brain following damage. After surgery, split-brain patients very slowly become able to make simple utterances using only their right hemisphere, an ability that takes around 15 years to develop. Mike's explanation is that there is a 'naturalistic' experiment going on all the time in normal brains – the right hemisphere sees normally and orchestrates responses with the left hand. 'Pretty soon', says Mike, 'it seems that there is a latent speech centre there that can be activated'.

But are we right to assign clear-cut functions to the two hemispheres – speech to the left and nonverbal skills to the right? Once again, it seems that this breakdown of roles is simplistic. Damage to part of the right hemisphere destroys the natural melody of speech. People with this problem can still speak, but they have a completely flat, emotionless tone of voice. Likewise, damage to another part of the right hemisphere makes it impossible to *hear* the emotional tone in voices. Hence, language is not processed exclusively in the left hemisphere.

One study suggests that the breakdown of roles between the hemispheres is not verbal versus non-verbal, but analytical versus emotional. The guinea pigs in this study were music students. At the start of their university course they were given a standard test to see which side of the brain was dominant for music. As in most people, it turned out to be the right. After three years of study, however, dominance had moved to the left hemisphere. The explanation for the switch was that the students' brains had learned to process music in a more analytical way – they now approached music with the same degree of analysis that most of use to learn a language.

But the important conclusion to draw is not that one hemisphere handles analysis and the other emotion, nor that one handles language and the other nonverbal skills. Rather, the finding suggests that the difference between the left and right hemispheres is one of degree rather than of absolute distinction. In other words, we should be wary of ascribing rigid functions to particular parts of the brain. This habit of matching function to lobe is an old one in neuroscience, borne largely out of the rationale of inferring function from dysfunction. If damage to Broca's area causes language impairments, the traditional argument runs, then Broca's area must be the centre for language. The fallacy in this conclusion is obvious if you imagine the same logic applied to a radio – if you took a valve out of a radio and it started to howl, that would not mean that the function of the valve was to inhibit howling.

Compelling evidence that there is no such thing as a 'language centre' comes from fascinating studies of speech during brain surgery. Neurosurgeon George A. Ojemann of the University of Washington, Seattle, used an electrode to stimulate the brains of his conscious patients, much as Henry did with Sarah in chapter one. George asked his patients to speak as he applied the electrode to different parts of the cortex. At certain sites in the left hemisphere, speech became slurred; then, at one place, it stopped. Remarkably, George found that the sites where speech stopped varied greatly in location from one person to the next.

George has pioneered another approach, one that enables him to listen in on neurons at work. He uses electrodes to pick up the fleeting electrical signals that buzz across the cortex while a person talks, allowing George to map which areas are active. The picture that emerges is far more complex than one of simple monopoly of language by the left hemisphere. It turns out that large numbers of neurons in the right hemisphere are active as well.

So what could be happening in the right hemisphere during language? The answer, according to George, lies in what might be happening *between* the hemispheres. The two sides of the brain are clearly doing different things, but the differences are subtle. The left hemisphere, he thinks, is responsible for word-finding and understanding, whereas the right may be involved in background work – the emotional colouring, the broader nuances of meaning. The result is that our two hemispheres enable us to see the world at more than one level, to see detail and the bigger picture at once. Perhaps this is why damage to the right hemisphere can flatten a person's tone of voice – their language abilities remain intact, but the emotional content is destroyed. It does, after all, make no more sense to speak of a hemisphere for a particular function than it does to speak of a 'centre' or a neurotransmitter for any one distinct function.

George sums up: 'There is some evidence that the way you respond verbally to people's facial expressions is something that depends on both the right and the left. The left sort of provides the output, but the right provides the perception of the facial expression and the emotional tone of it.'

If George is right, then language – far from being the preserve of a discrete language centre – is not even a single function within the brain. It is handled in a distributed fashion by many regions working in parallel. Perhaps we evolved the gift of language by a change in the balance of power between those regions.

Language has made us what we are. It freed us from stereotyped gestures and allowed us to use symbols to think metaphorically, to see one thing in terms of something else and to use not just words but art to represent complex relationships, which, in turn, have inspired innovative ideas. And these ideas have now been passed from one generation to the next, freeing us from the need to start from scratch with each new generation, unlike our primate cousins, who have not shown the same sharp technological learning curve.

Above all, however, language gives us a symbol for something that normally does not make inroads into our senses, simply because it is always there: one's self. As soon as we have a simple word for ourselves then we can inter-relate the self in a context. We can become self-conscious. This self-consciousness, combined with the ability to escape the here and now, is surely what really distinguishes us from almost all other animals, as well as from seemingly inhuman human infants. Language is merely the means, not the end. The critical issue is that we seem to differ from other animals and human infants in our ability to be massively self-conscious – introspective – and to filter, or even disregard, the sensual press of the present moment in favour of some inner cogitation centring around a self. But we must not forget that these are sophisticated optional extras that are relatively new in evolution. Before self-consciousness and imagination, and even more of a greater mystery to solve, is consciousness itself.

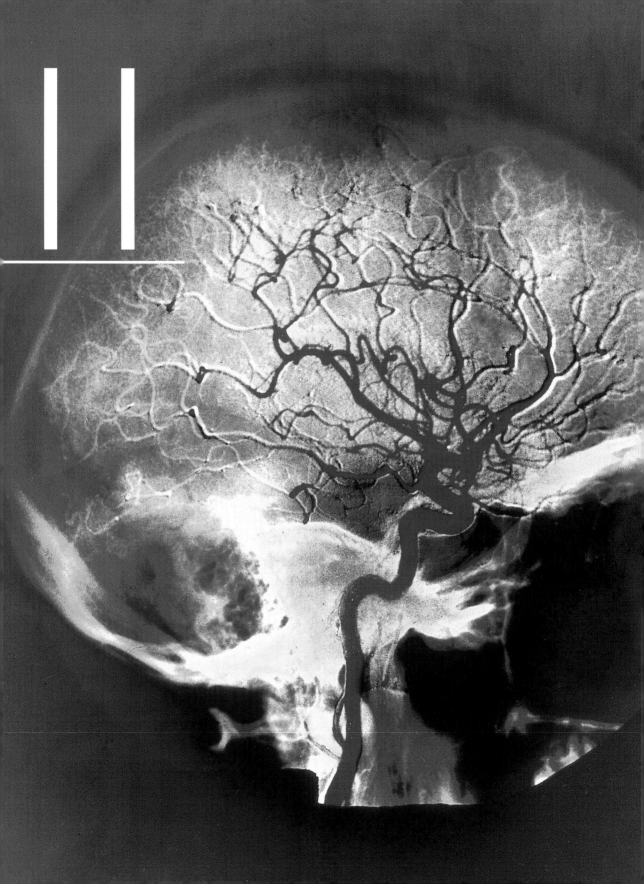

THE RIDDLE OF CONSCIOUSNESS

Ninety-five football fans were crushed to death in the 1989 Hillsborough tragedy, when crowds surging through the UK's Hillsborough football stadium became trapped against a barrier. Tony Bland was one of the survivors, but he was left with severe brain damage and never recovered consciousness. Tony spent about four years in what doctors call 'persistent vegetative state' until the government finally granted his parents permission to withhold feeding and let Tony die.

Although Tony had survived, he had lost that all-important brain function that makes life worth living – consciousness. Consciousness is that world that only you can access, and that makes you special and different to those around you. This first-person, personal world is one of the last great mysteries of science. We all know what it is, yet it defies definition.

We can think of consciousness as the state we enter each morning when we wake up, and depart when we go to sleep, yet we know there is more to consciousness than this. Tony Bland could 'wake up', but he had

◀◀ An artificially coloured angiogram of the skull showing arteries in the brain. It is through the circulation of blood that chemicals are brought from the brain to the body.

		PERSISTENT VEGETATIVE STATE

PERSISTENT VEGETATIVE STATE

Within about two weeks, a patient in a coma will usually have recovered, died or entered persistent vegetative state (PVS). PVS is similar to coma in that the patient lacks any apparent sign of self-awareness or consciousness and is unable to move. The important difference, however, is that in PVS the normal sleep–wake cycle is intact, the eyes opening and closing as normal. Someone in PVS can be 'woken up', although when awake they don't respond to stimuli and they seem incapable of higher thought processes. Moreover, their brain wave pattern, as revealed by EEG, is abnormal. A second, as yet unidentified 'cognitive' component in the brain clearly vital for consciousness, is missing.

This component would be related to the higher brain functions – involving brain regions beyond the brainstem. PVS patients may give the illusion of consciousness – they can breathe unassisted, chew, swallow, focus the gaze of their eyes, and their pupils respond partially to light. They even smile or scream occasionally and make random movements of the head or limbs. In this respect, PVS is very different from the distressing 'locked-in syndrome', where no movement is possible but the patient has normal brain function and is fully conscious. A current estimate of the prevalence of PVS, in the USA, is some 10,000 to 25,000 adults and 4,000 to 10,000 children.

lost the ability to experience the world. And even when we are awake, there seem to be different degrees of consciousness. The brain can achieve a great deal *subconsciously*, when we do something without being fully aware of it. Take driving, for example. You can drive for miles without consciously thinking about when to press the clutch, how to change gear, or how much to turn the steering wheel. There you are, free to roam in your own private inner world, free to plan what to eat for dinner or who to go out with for the evening. Yet a child running into the road will snap you out of your reverie, suddenly focusing your consciousness on the task of stopping the car. So what determines what you are conscious of, and what remains lurking in the subconscious? It feels as if there's a control centre in the head, an executive that selects where to focus consciousness. Of course, the idea of such a control centre doesn't solve the problem, it just miniaturizes it. Even if there was some kind of brain within a brain, we would still be at a loss to explain how it worked.

Can we discover anything about consciousness by comparing what happens in the brain during a subconscious action to what happens when a person is in conscious control of their actions? A simple enough experiment is to scan the brain while someone is learning a tricky skill, such as a specific sequence of finger taps. It turns out that, while the person is focusing their conscious mind on learning the task, vast areas of the brain light up, particularly the prefrontal cortex. But once the skill has been practised enough to become second nature, many of these areas appear to shut down. The process becomes automatic, and the subconscious autopilot (which seems to involve the cerebellum) takes over. So is the prefrontal cortex the centre for consciousness?

Let's return to Graham Young, whose rare visual disorder – blindsight – has made him unaware of objects on his left-hand side. Graham has lost consciousness of only part of his visual field, an impairment that has made him a superstar in the field of consciousness research. He claims he is guessing when asked to point at objects on his left, yet his guesses are generally correct, so his brain must be processing vision subconsciously. Graham's plight has given experimenters an opportunity to address the involvement, or otherwise, of the prefrontal cortex in the generation of consciousness.

Although Graham has lost part of his consciousness, imaging studies reveal that his prefrontal cortex is in good working order and entirely intact. So the prefrontal cortex cannot be the centre for consciousness – if so it would have reflected in some way Graham's impairment. In Graham's case what is lost is part of the pathway projecting to the prefrontal cortex, a part of the visual cortex involved in the first stages of visual processing. This observation suggests that it is no particular single region that is important, but the connections between them need to be intact so that the brain can function cohesively. When it comes to consciousness and brain function, the whole is clearly more than the sum of the parts.

But how much do such experiments, highlighting as they do the transition of subconscious to conscious, really tell us about consciousness in general? After all, the observations on Graham's condition involve only one type of consciousness – vision – yet most importantly, any blindsight patient is actually fully conscious all the time. As American philosopher John Searle has pointed out, there is surely a big difference between modifications to an existing conscious state and no consciousness at all.

'Well, I think the study of blindsight is absolutely fascinating', says Searle. 'However, I think it's a mistake to suppose you're gonna crack the problem of consciousness with blindsight, that you'll find the difference between consciousness and unconsciousness by looking at the difference between the normally sighted and the blindsighted person. In the case of the blindsight studies, they are always done on conscious subjects, that is the guy who exhibits blindsight is conscious already, and what we want to know is: what's the difference between the conscious and unconscious brain?'

In blindsight research and other studies of consciousness, the idea is to report consciousness of a display such as a grid of alternating colours that pulses at a certain frequency while the subject lies in a brain scanner. Perhaps the most we can learn from such experiments is which brain regions are active in certain situations. We might learn how the brain changes for consciousness of one thing compared to consciousness of something else; but we will not learn how consciousness itself is generated in the brain in the first place – how a subjective, inner world is generated from a lump of neuronal sludge.

'Blindsight is an example of the building-block approach', says Searle. 'Our conscious field is made up of these little building blocks – there's one for colour, one for shape, and all of these add up to vision. The idea is that if you find out how the brain builds one building block, you've cracked the whole thing. The building-block approach would predict that if you triggered the building block for any particular experience, then an otherwise unconscious subject would have that experience. If you triggered the building block for red, the unconscious patient would suddenly have a conscious experience of red. I don't think that's possible.

▲ Trafalgar Square photographed in 1870. The camera in those days was too slow to capture images of moving people, but some of their shadows can be seen.

I think you can only have the experience of red perception if you've already got a conscious subject, if you've already got a unified field of consciousness.'

The problem is how to capture the 'unified field of consciousness', whereby the brain is somehow different from its unconscious state. If you look at early Victorian photographs of cities in broad daylight, remarkably there are no people – the buildings are captured with stark clarity and detail, yet the streets are deserted. The reason is that the exposure time of those early cameras was too long to capture the citizens going about their business, or even loitering in the street. So it might be with current imaging technology. We see the structures of the brain, yet the techniques might be too slow to capture something else, some vital process that somehow unifies all the brain regions and is fast and transient enough to match up to a moment of global consciousness.

Perhaps a good way of starting to tackle the riddle of consciousness is to investigate what happens when consciousness is absent altogether. The most obvious and ubiquitous example of such a state is sleep.

We do not know if dinosaurs slept, but all birds and mammals for the last 200 million years have displayed some type of sleep pattern. Sleep is a fundamental function of any reasonably elaborate brain. In contrast to sleep as we mammals know it, the rest periods of insects are far less complex, yet nonetheless distinct as a different state. For example, in honeybees electrical activity changes in the brain during nocturnal resting periods. So, surely, if even insects can lose and regain something, that something is a primitive form of consciousness. Sleep can be determined by a range of signs: a slowing of body metabolism, relative immobility and, of course, loss of consciousness (except in dreaming, with which we shall deal separately).

We can study sleep in humans by placing electrodes on a person's scalp and monitoring their brain waves during the night on an EEG. The resulting pattern reveals four distinct stages of dreamless sleep, each with a characteristic brain wave. When we fall asleep we descend rapidly through the four stages, from the lightest to the deepest form of sleep. Throughout the night, we gradually surface and descend again, cycling through the four stages four or five times. Studies of sleep suggest, therefore, that unconsciousness comes in degrees. If correct, it seems reasonable to suppose that consciousness might be similarly graded – you can have more or less of it at different times. A similar idea emerges from studies of another form of unconsciousness, anaesthesia.

At Addenbrooke's Hospital in Cambridge, UK, Gareth Jones and his colleagues have been giving small doses of anaesthetic to student volunteers to try and work out what disappears in the brain when we become unconscious. Gareth has found that there is no one area of the brain that is switched off when we are anaesthetized; rather, the effect of anaesthetics is somehow to deaden the whole brain by suppressing action potentials everywhere. Yet this deadening is not immediate. There are four recognized stages of anaesthesia (analgesia, excitement, surgical anaesthesia and depression of breathing centres). As a person progresses through the stages they lose consciousness gradually, rather like turning down a dimmer switch. If consciousness is not some magic switch that

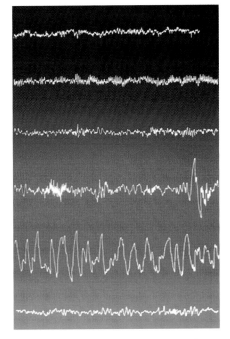

▲ EEG recording of brain waves. From top to bottom:
1 The brain when awake
2 Awake but with eyes closed
3 Falling asleep
4 Early sleep
5 Standard sleep pattern
6 Rapid eye movement (REM) sleep, when dreaming occurs.

flicks on or off, but instead something quantitative, something you can have more or less of, it might be easier to find corresponding ebbing and flowing situations in the brain.

The trouble is that no one seems to understand just how anaesthetics work. Irrespective of their differing molecular structures, they all appear to have the same common effect – deformation of neuron membranes. We saw in chapter two that cell membranes play an important part in allowing ions to traffic in and out of the cell, so creating the electrical signal – the action potential – that forms the basis of all brain functions. By changing the configuration and composition of the membrane, anaesthetics reduce the ease with which a cell fires off its electrical signals. This suppression occurs all over the brain, not in one particular region.

So any theory of consciousness cannot rely on particular brain regions or particular cells. There is no 'centre' for consciousness – after all, that would mean we have a complete mini-brain in a brain. Likewise, there is no single area that, when suppressed, triggers unconsciousness. At the other end of the scale, we know that consciousness cannot reside piecemeal in individual neurons, since some neurons continue to work during anaesthesia or sleep.

At some more global level of organization, then, beyond a single cell and beyond any rigidly demarcated edifice in the brain, we might imagine an assembly of neurons so large that it does not respect any particular anatomical boundary, and at the same time one that is fast and evanescent enough to permit the formation of another, different and dominant assembly a moment later.

We have already seen that consciousness is not all-or-none but comes in degrees. Indeed, most of us would accept that we have different degrees of consciousness at different times – being on a mountain top or meditating arguably entails a greater consciousness than staring at a TV. This sliding scale of consciousness could not easily be accommodated in a fixed brain structure, but would depend on a more widespread, higher-order brain state that could cater for the contraction and expansion of our subjective experiences. The greater the assembly size at any one time, perhaps the greater the level of consciousness.

The idea that consciousness comes in degrees, and that it requires large assemblies of neurons to achieve even the faintest level of consciousness, could explain recent findings that people in persistent vegetative state show islands of activity in the cortex. This finding, in itself, is yet further evidence that the cortex alone cannot be the seat of consciousness; more importantly, it also indicates that no particular circuit between regions could alone act as the basis of consciousness. Even when different areas are active, a state of unconsciousness still prevails. What is missing in the PVS patients is coordinated activity across multiple brain regions – or, as I would have it, large-enough neuron assemblies.

Happily, I am not the only scientist who thinks that this idea of neuron assemblies might be the best indicator of degree of consciousness. German neuroscientist Hans Flohr suggests that as a person succumbs to anaesthesia the neuron assemblies get smaller, and the degree of consciousness correspondingly less. Hans is currently working out how this scenario might actually come to pass. His experiments suggest that neuron assemblies might form through a particular type of neurotransmitter: glutamate. Glutamate binds to a receptor known as NMDA and, in doing so, makes the target cell easier to excite, and hence easier to recruit into a transient assembly. Hans believes that this sequence of events is key for the formation of large-enough assemblies in the conscious brain. He suggests that the NMDA receptor is the very site that, ultimately, is critically affected by anaesthetics, albeit in a variety of ways. Accordingly, he has shown that a range of different anaesthetics reduce activity of the NMDA system consistently.

One such anaesthetic is the drug ketamine, which blocks the glutamate receptor. Ketamine is perhaps better known as a drug of abuse, taken for its hallucinogenic qualities. Its effects are being studied in the more rigorous regime of the laboratory by Swiss neuroscientist Franz Vollenweider, who has been giving volunteers ketamine and asking what they feel. At high doses the drug merely causes loss of consciousness, perhaps because it prevents any appreciable size of neuron assembly from forming. At weaker doses, however, I would suggest that assemblies could still form, albeit abnormally small ones. Interestingly enough, it is at these doses that the subjects report hallucinations.

An idea that has intrigued me – and it is just an idea – is that small neuron assemblies might be the basis of a type of consciousness that does not access the rich connectivity we have seen makes up an adult mind. Rather, this type of consciousness might deal solely with the present, reacting to the outside world in a literal and passive way. In this state of reality, the outside world is not placed against all the checks and balances that one has learnt gradually through life – hence the lack of logic and the suggestibility that is so characteristic of hallucinations. Drugs such as ketamine would not be the only means of achieving this form of consciousness.

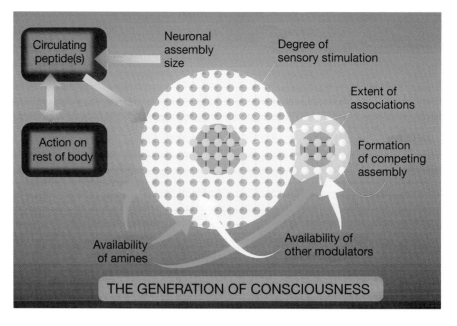

▶ Greenfield's theory of consciousness. The two sets of concentric circles represent two transient assemblies of tens of millions of brain cells: the largest assembly will dominate at any one moment in the brain, and determine that moment of consciousness. The degree to which cells are recruited, and hence the degree of consciousness, will be determined by a variety of factors, such as vigour of sensory inputs, pre-existing connections ('associations'), and degree of competition ('distractions'), as shown by the smaller assemblies starting to form. Signature chemicals, peptides, will be released from the transient assembly. The type, number and concentrations of these peptides will thus represent the salient assembly in the brain, and convey that information to the rest of the body, via the circulation. In turn peptides released from the immune system and the vital organs will modify the working assembly of neurons, as will other chemicals such as hormones and the 'amines' that are released in relation to arousal. Consciousness is therefore dependent on the whole body working cohesively.

In childhood, too, because the connectivity is not well-developed, assemblies of neurons would perforce be small, and the corresponding consciousness one that is literally and vividly trapped in the present moment. Children and drug-takers have one thing in common: unlike adults, they ride an emotional roller coaster. As we saw earlier, strong emotional states are characterized by an awareness of only the present and an absence of thought or logic. So it seems to me an attractive idea that emotions are the most basic form of consciousness and, in turn, may be related to small assemblies of neurons in the brain.

If so, then we would not need to be a child, nor dependent on drugs, to achieve a small assembly state. At certain times in our lives, even if we are not in an emotionally charged fight-or-flight scenario, it might be hard to recruit a very large assembly in our brains. This time the reason would be that input from the outside was minimal, and not therefore exciting a large number of cells; a further reason would be because of the surging of the different brain chemicals such as acetylcholine, noradrenaline, dopamine and serotonin at certain times in our 24-hour sleep–wake cycle, all of which would differentially influence how easily our brain circuits could connect up.

As well as the four stages of sleep through which we cycle several times a night, there is also another type of sleep, one that is totally different from the rest. During this stage of sleep our eyes move rapidly backwards and forwards – hence its name, REM (rapid eye movement)

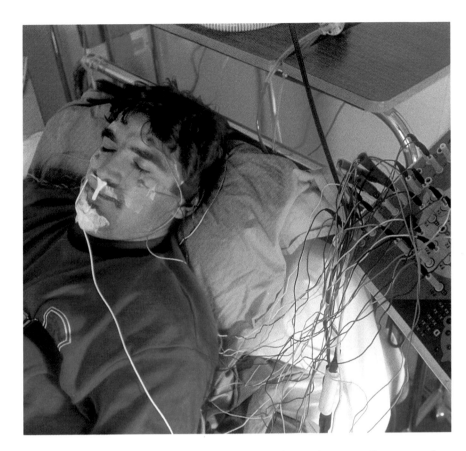

◀ A man wired up to somniography equipment used to study sleep. Electrodes are attached to a patient's scalp to record electrical activity in dfferent parts of the brain.

sleep. If people are awakened during REM sleep, they usually report that they have been dreaming.

REM sleep actually resembles wakefulness in that the brain stops manufacturing the housekeeping chemicals, important for cell maintenance, that are made during dreamless sleep. Indeed, as far as imaging studies and EEG traces can tell, activity in the brain during sleep and wakefulness are indistinguishable. And although we may toss and turn in dreamless sleep, in REM sleep our muscles become paralysed to stop us acting out our dreams. Only our eyes dart about, perhaps looking at the images moving about in the dream world. So dreaming, then, is really a form of consciousness. Our dreams may be imagined, but they are as real to the brain as the world we see when we are awake.

Not all animals dream. Reptiles do not show REM sleep at all, and birds do so only occasionally, mostly as hatchlings. On the other hand, all mammals – at least according to their EEGs – seem capable of dreaming throughout life. In an average night's sleep of seven or eight hours, adult humans spend a total of 90–120 minutes dreaming. Children dream more than adults, and the amount of dream time gradually falls with increasing age. Incredibly, a 26-week-old human foetus spends all its time in REM sleep.

The purpose of dreams is something of a puzzle. A common idea is that dreaming helps us to sort out the experiences of the day, but if this theory is correct, what experiences is a 26-week-old foetus consolidating in its never-ending dreams? A more down-to-earth idea is that dreams are a consequence of the way an immature brain functions. In a young brain, the connections for any potential assembly are still forming. By contrast, in the brain of an adult, connections are well-established – although during a dream these connections are insufficiently activated due to the absence of stimulation from the senses. Although the causes are different, the net effect would be the same – abnormally small neuron assemblies, producing a crude level of consciousness where raw feeling, not logical thinking, is the dominant feature.

If true, this idea could have another interesting implication – perhaps the illogical consciousness of schizophrenia is similar to the conscious-ness of dreams. It might be the case that a central problem in schizophre-nia is reversion to small neuron assemblies, leading to a dream-like – or, perhaps more accurately, nightmarish – view of the world where logic no longer works. In the case of the schizophrenic's brain, however, the cause of the impairment is not lack of neuronal connections nor absence of sen-sory input, but an imbalance of dopamine. And transmitters, as we have seen, would also be a factor in the ease of assembly formation.

In my own view, dreaming represents the lowest end of a continuum of consciousness, low on logic and high on feeling, resulting from small assemblies of brain cells driven by residual brain activity. Viewed in this way, we might place nonhuman consciousness, too, on a sliding scale. Perhaps, as with the developing foetus, consciousness increases with brain size across the animal kingdom. If less conscious states are charac-terized by smaller assemblies of neurons, we would expect that animal consciousness is one of living in the present moment, where raw feelings hold sway over reasoning, memory or planning.

In humans – whether in childhood, dreaming, schizophrenia or states of intense emotion – small neuron assemblies create a consciousness dif-ferent from our normal, thoughtful state. Unable to reflect on the past or speculate on the future, we become trapped in the present moment, the passive recipient of tastes, colours, smells and sounds stripped of person-alized memories and significance. Perhaps pleasure is such a state, one in which we attempt to focus only on the present and block out worrying thoughts about the mortgage, the job loss, the future of one's wayward child. The passion of the moment – whether we are caught up in music, dancing, sport or love – stops more complex neural assemblies from impinging on our consciousness. Perhaps it is by manipulating such neural assemblies that some people are able, literally, to accomplish extraordinary feats of mind over matter.

US circus star Jim Rose makes his living by doing exactly that. He can hang weights from his tongue, swing irons from his nipples, and push his face into broken glass. Jim's secret, it seems, is to establish a state of mind that enables him to keep pain out of his consciousness.

He describes his thought process during a stunt: 'where I like to go is into a nice warm-watered pool, and I like to be right up to my neck and everything's feeling fine, everything's already nice and warm and relaxed. No, I don't, I don't feel the pain. I've worked it out to where it's mild discomfort.' But things are different if Jim is mentally unprepared. 'If I stub my toe on my bed at night, I'm going to scream just like you, because I didn't expect it and I didn't put myself in the warm water.'

My explanation is that Jim is creating, in advance, a large and dominant neuronal assembly that prevents any competing assembly from forming. If the sensation of pain happened to require a particularly large assembly, then restricting its growth would curtail that sensation. Indeed, there is evidence that pain does depend on large assemblies. After all, we describe painful stimulation in terms of other associations – pricking, stabbing, burning and so on. Perhaps painkillers such as morphine work by preventing large assemblies from forming, producing dream-like side effects for the same reason.

▲ Mind over matter. Performer Mr Lifto, in Jim Rose's circus, displaying extraordinary feats: somehow he keeps pain at bay.

Understanding consciousness in terms of these transient assemblies may help us understand how the brain creates the wealth of mental experiences that form our lives, from simple thoughts to hallucinations, from anaesthetics to feats of mind over matter. But there is another important consideration – as yet, these transient assemblies are hypothetical phenomena since, as with the old Victorian photographs, current techniques would be too slow to capture them. Nonetheless, some evidence from animal brains, reported by Amiram Grinwald in Israel, reveals that assemblies of brain cells can actually arise for brief periods of time. In response to a flash of light, for example, Amiram has found that as many as 10 million brain cells can be temporarily activated in just under a quarter of a second. Although too fast to be captured by a brain scanner, this activity is quite slow in brain terms – these neuron assemblies spreading across the brain take time to build up. If growth of an assembly of neurons does indeed create a moment of consciousness, then there must be a delay between the arrival of a stimulus and its subsequent entry into consciousness.

Just such a delay was first discovered in the 1960s by physiologist Benjamin Libet. Libet started out with a simple query: how long does it take for the brain to register a pinprick on the hand? He worked with patients undergoing brain surgery while awake, rather as we saw with Sarah in chapter one. In these particular patients, however, Libet measured brain activity. What he found surprised him. When he pricked a person's hand, it took a mere 20 milliseconds for the signal to reach the brain, but a full half second (500 milliseconds) before the person started to report that they had felt anything. So, although the brain can unconsciously process the information almost immediately, it takes a full 500 milliseconds for 'you' to become aware of what is going on.

Half a second is a long time in the world of the brain, where an action potential can be over and done with in less than 2 milliseconds. This means that much of what we do, and what we think of as skilled, deliberate and decisive, must be happening too quickly for us to be conscious of it.

A top-class tennis player has a serve of up to 120 mph (193 kph) – once the ball has left the racket, his or her opponent has under 400 milliseconds to work out where the ball is going to land. The decision about how to return the shot has to be made subconsciously. Amazingly, when a player returns the ball they are not even consciously aware that the serve has started. Vaughan, a tournament-level player, tried to explain:

'Well, the decision's really an instinct. You don't decide where you're gonna or when you're gonna hit the ball. It's just a matter of deciding where you're gonna hit it. And you don't really know when you're gonna hit the ball until the ball's come.'

None of the expert players I interviewed claimed to make a conscious decision about how to return a fast serve. Yet such a response is far more than just a reflex – it involves thinking strategically about exactly where to place the shot. Returning a tennis ball is a complex process – all done entirely subconsciously.

British psychologist Jeffrey Gray wonders why, if the brain can perform so well subconsciously, it bothers being conscious at all. My view is that there is more to consciousness than simple physical skills such as returning a tennis ball. Surely the most important thing about consciousness is that it allows us to have an inner life. Hand–eye coordination on the tennis court is small beer indeed compared to the feel of sun on your face, or any other of the wealth of sensations and thoughts that make up our lives. We might be able to play tennis without consciousness kicking in, but living life unconsciously would not be a rewarding or worthwhile experience – as the legal decision on Tony Bland stands witness.

If the brain on the tennis court can make decisions without conscious intervention, of what else might it be capable? We rely on conscious decisions every day, whether choosing what to have for lunch or decoding which insurance policy to go for. But who is really in control here, you or your subconscious brain? It certainly feels as if the conscious self makes the decisions that sets the body in motion, but can we find out for sure? Benjamin Libet decided to try.

▲ Tennis star Pat Rafter serving. The speed with which great players automatically react to fast balls demonstrates the brain's ability to work subconsciously.

Libet explored what happens in the brain when we decide to move – which comes first, the conscious decision or the unconscious motor action? His experiment was recently repeated by physiologist Patrick Haggard of University College, London. The setup is very basic, as I discovered when I became Patrick's guinea pig. Electrodes on my scalp monitored electrical activity in my motor cortex, the part of the cortex concerned with generating movement. All I had to do, at any time of my own choosing, was to press a button and say exactly when I had the desire to press it.

Libet had expected the conscious desire to move to come before the motor areas of the brain started working, but he found exactly the opposite. The decision to act comes nearly a whole second *after* the motor areas have started preparing themselves for action. Your brain has already subconsciously made the decision to move, and 'you' only find out about it once the process is in action!

The implications of this finding are staggering. If the intention to do something comes after the brain has already decided to do it, if the brain makes its decision before 'you' do, then our behaviour is guided not by free will but by subconscious processing. The feeling of 'you' – the individual in your head – could well be the most impressive trick the brain plays. Somehow, the brain creates the illusion of a conscious self in control of its actions, while the true controlling force is the subconscious.

The idea that there is some sinister biological process within the brain that manipulates us – what American philosopher Dan Dennett calls the 'selfy self' – is certainly an uncomfortable notion. But if our conscious thoughts come from subconscious processes, *then that has to be the case.* Consciousness is a product of the subconscious. In objective brain terms, the feeling of consciously being in control is a neurophysiological con trick. This idea is perhaps not as heretical is it may sound. After all, all our thoughts and actions derive from the activity of the brain, whether subconscious or conscious. These are not two sparring entities – self and brain – but part of an integrated whole. All 'you' consist of is a brain, albeit one personalized by a unique trajectory through life.

If consciousness and free will are mere illusions, where does that leave personal responsibility and accountability? Does it mean that we can't hold criminals responsible for their actions because their behaviour arises from involuntary subconscious forces? As Deborah Denno, Professor of Law at Fordham University, New York, points out, the trouble with this view of the brain is that we have to throw out the notion of the 'person inside' – the individual that makes decisions and is responsible for its own actions. Yet, if our behaviour can be reduced to shrinking and expanding assemblies of neurons, whether it be a conscious or subconscious process, then to what extent can our behaviour be said to be free? If our actions are predetermined by the way our brain is connected and, in turn, by how those micro-circuits of connections function, then how can 'you' be held responsible for your actions? For example, the *crime passionnel* in France, in which mitigation for a crime is merely extreme emotion, could be interpreted as an extreme example of a small neuron assembly in the brain, one in which logic and reason play little part.

Despite all that we know about networks of nerve cells, neurotransmitters, and subconscious processes, it still *feels* as if there is a conscious inner self that is free to make choices. Our ability to reflect on this inner self highlights an important quality of human consciousness: we are not only conscious of the world and what's going on around us, we are also conscious of ourselves – we are self-aware. Self-awareness is very rare in the animal kingdom; only apes share with us the ability to recognize themselves in a mirror. But even they show no signs of being capable of insight, of introspection, of contemplating their future. Perhaps self-consciousness is an intense form of consciousness, requiring exceptionally large neuron assemblies and sophisticated brains. In my view, it is a natural extension of the raw, primitive forms of consciousness experienced by animals and young children. As adults, self-consciousness is our normal state, although we sometimes literally 'let ourselves go', abandoning self-consciousness and living for the moment, as in times of intense emotion, dreams, drug-induced states or, perhaps, schizophrenia.

As brain-imaging techniques get ever more sophisticated, and able to operate over increasingly tiny timescales, it may become possible to see at first hand the flickering assemblies of neurons that correlate to different forms of consciousness. Such advances in technology would allow us to assess the theory of transient assemblies by generating testable predictions. For instance, we might predict that certain kinds of assembly would appear in children, drug-takers or schizophrenics. And we might be able to correlate assembly types to variations in brain chemistry, sensory stimulation or age. But such speculation takes us into the realm of the future, and that is the subject of the next and final chapter.

12

THE BRAIN OF THE FUTURE

Sarah has made a good recovery from the operation described at the beginning of this book. Her brain, exposed to the scrutiny of the surgical team, myself, and the unblinking fish-eye of the TV camera, has remained resolutely her own, a personal place that no one else can access directly. Like the rest of us, Sarah now faces a new millennium in which science and technology will impact on our lives as never before. In the 100 or so millennia since the first appearance of *Homo sapiens*, humans have taken their private mental worlds, their individuality, for granted. But now the opportunity to meddle with the brain is greater than ever before. That most personal part of our anatomy may be in line for a dramatic makeover – both indirectly, through a change in the way we use our brains, and directly, through precision surgery, genetic manipulation and drugs. What are the implications for our future as individuals? Will we still be individual at all?

One of the greatest threats to the brain of the future is the disturbing rise in neurodegenerative diseases, such as Alzheimer's and Parkinson's diseases. The most urgent issue to address, then, is whether we will be prepared to tackle these devastating conditions. As we saw in earlier chapters, the only currently effective drugs are palliative – they relieve the symptoms without affecting the cause. Yet work is already afoot to identify why it is that only certain cells in the brain should die slowly and inexorably, as happens in neurodegeneration. If we can find the common factor that makes these neurons special, we may hold the key to a cure.

Recent research has shown that the neurons lost in Alzheimer's and Parkinson's diseases are strikingly different from other brain cells. Remarkably, they retain the ability to regrow; all other neurons lose this ability once they mature. One idea that I find compelling is that neurodegeneration may an aberrant form of growth that is somehow toxic to the cells – in retaining the ability to regrow, the neurons shoot themselves in

◀◀ DNA gel analysis. A researcher in protective clothing studies DNA fragments under ultraviolet light. The fragments have been stained to show up as fluorescent violet bands on the gel.

the foot. If we could find out more about this mechanism, we might be able to arrest neurodegeneration upstream, targeting the cause rather than the symptoms. We could then offer patients, if not a reversal of their memory loss or movement difficulties, then at least stabilization of the condition, so that no further deterioration occurred. And if a reliable diagnostic screen could be developed, perhaps with a blood test, then the prospect of effective prevention would be on the horizon.

Research into Alzheimer's and Parkinson's diseases may lead to new ways of engineering the brains of the elderly. The Human Genome Project holds the prospect of revolutionizing medicine for those at the other end of the age spectrum – the newly conceived. Once the Project is complete, scientists will have a detailed map of every working gene in the human body. Inevitably, the debate over how we use that information will become even more ferocious than it is today.

Ever since Dolly the sheep was cloned in 1997, fears, hopes and, above all, imaginations have run riot. The last decade has witnessed an explosion in knowledge and know-how that would have been impossible to envisage when the basics of gene function were first discovered in the middle of the 20th century. Already, medical techniques based on genetics are routine. One such technique is genetic screening, in which parents at risk of having children with a genetic disease are screened for the aberrant gene and, if necessary, given counselling on the implications of starting a family. If a pregnancy is underway, parents have the option of deciding to terminate if the foetus tests positive for the disease.

Genetic screening is carried out only to detect diseases caused by a single faulty gene, such as the lung disease cystic fibrosis. If scientists could somehow repair or replace the faulty gene in such disorders, they might be able to cure them. This is the rationale behind gene therapy, an experimental technology that has so far failed to live up to its promises. The challenge in gene therapy is to supply millions of cells at once with a working copy of the gene in question, but this has proved far from easy to achieve.

At the moment the best established gene disorder based specifically in the brain is Huntington's chorea, the disorder of movement that causes wild, involuntary flinging movements of the limbs. Sadly, other brain diseases do not have such a clear genetic provenance. For example, scientists have found at least six genes that seem to increase the risk of the most common form of Alzheimer's disease, but none of these genes lead inevitably to the disease and none are essential. Genetic screening alone, therefore, cannot determine whether someone will go on to develop the disorder; at best, it can give an idea of the likelihood. And it seems unlikely that gene therapy will offer much hope of a cure. In any case, diseases such as Alzheimer's, which appear only in later life, do not prevent a person from leading a happy and fulfilled one; avoiding the risk of a neurodegenerative disorder in old age is no reason not to live in the first place, and hence poor justification for terminating a pregnancy.

Perhaps a more fruitful application of genetics will be in the study of the genes that control development in the human embryo. Such research might shed light on the aberrant growth mechanism that seems to trigger neurodegeneration in adult cells. Moreover, it might one day enable scientists to reprogramme genetically adult brain cells so that they grow, for instance, into the dopamine-producing neurons that are lacking in Parkinson's disease.

But gene science in the future will be able to do much more than just patch up damaged tissue and help cure diseases. Among its more radical possible applications are techniques such as cloning and production of 'designer babies'.

US physicist Richard Seed is a cloning enthusiast. 'Clones are going to be fun', he reckons. 'I can't wait to make two or three of my own self.' But a clone is not a copy of the self any more than an identical twin is – identical DNA does not mean identical brains. Take the infamous Kray twins, for example, the gangsters who terrorized the East End of London in the 1960s. They were identical twins, but one was gay and the other was straight. Even more than with naturally occurring identical twins, the cloned individual would be less like their donor, both physically and mentally. Growing up in different environments and separated by a generation, they could hardly have had a more different start to life.

Perhaps more sinister than cloning would be the use of genetic technology to create designer babies. It may become possible in the future to screen a batch of your own embryos to select the one with the best genetic profile. More sinister still, imagine a

▼ A 1930s poster, an early example of ideas for 'optimizing' the gene pool.

ONLY *HEALTHY* SEED MUST BE SOWN!

CHECK THE SEEDS OF HEREDITARY DISEASE AND UNFITNESS BY EUGENICS

world in which parents can pick which genes to put in, choosing genes for height, hair colour and nose shape, perhaps even trying to design an ideal mental portfolio. If – and it is a ludicrously big if – genes were discovered for intelligence (whatever that means), niceness, confidence and so on, who can say what might result from any particular combination. Genes do not have function locked up inside them any more than do neurotransmitters or brain cells. Instead, they work together in complex ways, producing a bewildering variety of effects depending on context, with only partially predictable results. Attempting to create a designer baby would be like opening a genetic Pandora's box.

The new millennium will bring us more choice – not just in the type of children we might have but in the lives we ourselves will live. The hugely increased possibilities for accessing information, for travel – both virtual and real – as well as the arguable economic and social benefits generated by a globalized culture and economy – are all offering thrilling new challenges. And there will be more time to benefit from these new opportunities: we will have more leisure because of longer life spans or because robots will be taking over the boring and dangerous stuff. Of course, there might also be the less happy spectre of redundancy and unemployment, but we are concerned here with how neuroscience, not economics, will affect the future. And as far as we are concerned the critical issue is that, finally, we will come face to face with ourselves and will have no excuse not to make the most of our lives. But then could come the crisis that has afflicted humans through the ages: what do I actually want out of my life? What is happiness? Why am I not satisfied? It is at this stage that the way ahead may seem so uncertain, the choices so wide and frightening. The individual may feel unable to rise to such challenges and attempt a more direct route to happiness: by taking drugs.

Humanity has been self-prescribing drugs to alter brain states since the dawn of time. But the use of drugs as tools to combat specific brain problems is really a hallmark of the 20th century. The use of both prescribed and proscribed drugs in this new century will, I predict, escalate.

As we saw in chapter eight, drugs that affect the brain often exert their effects by acting on specific neurotransmitters. Will our increasing understanding of neurotransmitters lead, therefore, to the development of useful new drugs? Herein lies the problem. We have already seen that there is no one-to-one matching of neurotransmitter to brain function, but rather that neurotransmitters work in complex multi-way seesaws with each other. Perhaps, once these multi-way seesaws are exhaustively

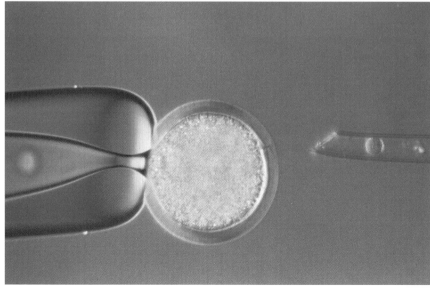

▲ The Human Genome Project. A researcher analysing the structure of DNA. This is a world-wide search to locate the genes responsible for inherited diseases.
◀ Part of the technique that produced the first cloned sheep, Dolly. Held by a pipette (left) is a sheep egg cell (centre), which has had its nucleus removed. On the right, a needle contains another cell about to be injected into the egg. A spark of electricity will fuse these cells into one, so that an embryo grows from this single cell.

understood, they will be programmed and processed with state-of-the-art IT accuracy – a 'polypharmacy' could then be developed that circumvents side effects because it exploits the multiple specific interactions and balances that together amount to the chemical signature of one particular brain region.

Recently, there has been a flurry of excitement over the so-called memory drugs, or 'smart drugs'. The idea runs that if these drugs can help arrest memory problems in Alzheimer's disease, perhaps they can boost the memory of a healthy brain. Once again, we must be cautious about claims for a magic bullet – the human brain is an unfathomably complex three-dimensional structure, not a two-dimensional bull's-eye. We can't explain memory in terms of a single chemical or a single mechanism on which a putative 'smart drug' could act. And, as ever, side effects would be inevitable. After all, cigarette smoking can improve mental concentration, but there would howls of protest if nicotine was promoted as a memory drug.

Most drugs that act on the brain work by simply blocking or boosting the classical process of synaptic transmission. Over recent years, we have seen enormous advances in our understanding of novel, nonclassical mechanisms that play more subtle roles in neuron operations, such as modulation, in which cells are put on 'red alert', primed to fire in response to a particular neurotransmitter. Now if a drug were designed to modify this modulation, then the effects would only be apparent under certain circumstances, when certain inputs to the cell were active. Then again, we must remember that whilst such drugs might cut down on the severity and number of side effects – they would be less of a chemical sledgehammer on the brain – they would still be working at the nuts-and-bolts level of the brain, not combating a net function such as memory.

Further virgin territory for the drug prospectors lies in the new classes of neurotransmitter that are being discovered, such as peptide neurotransmitters and the gas nitric oxide. As we saw in chapter two, peptides are sometimes present in the same neuron as classical neurotransmitters, but are released only when the cell is very excited. In such cases they broadcast a wider message than classical neurotransmitters, affecting several cells at once. In addition, peptides can be bilingual, functioning as hormones as well as neurotransmitters. As such, they operate over longer timescales and may have targets beyond the brain.

American neuropharmacologist Candace Pert will enter the scientific history books for her discovery of endorphins while still a humble PhD student. In a recent book she muses on whether there is a peptide for each emotion. Once again, beware the trap of over-reductionism of a complex brain state to the mere interstices of a molecular structure! On the other hand, the role of peptides in more generalized body functions, and indeed the interaction they might have with the immune system, present tantalizing possibilities for the development of future drugs. Already, for example, a new approach to depression has opened up with the discovery of a peptide called substance P.

Another promising molecule, dear to my own heart, is an enzyme called acetylcholinesterase (AChE). AChE is well known for its role of removing acetylcholine from synapses. However, evidence is accruing that this familiar chemical might have a completely different function, in modulating other types of brain cell. One such nonclassical role seems to be in development of the brain – amazingly, it appears that AChE can enhance neuron outgrowth in certain brain areas. Furthermore, there is a transient appearance of AChE in certain brain regions in development, that vanishes in maturity, but is recapitulated following damage. Clearly, AChE itself might hold promise for exploitation once its nonclassical, non-enzymatic actions are better understood. In addition, such compounds might inspire a different type of therapy for neurodegeneration. Remember the fascinating idea we explored earlier in this chapter: that neurodegeneration might be an aberrant form of development? If this proves correct, and if substances like AChE are pivotal in development, then perhaps it is they that are ultimately responsible, if activated inappropriately, for starting off the actual mechanics of neurodegeneration. And if so, they would pose a novel target for the development of anti-degeneration drugs.

This hypothesis, if true, would have enormous implications for another currently attractive idea for future treatments for neurodegeneration – actually to mimic trophic agents, a class of compounds that, like AChE, can enhance neuron survival in culture dishes and encourage rapid growth. The problem with these trophic factors, however, is that they are too large to cross the blood–brain barrier. The blood–brain barrier consists of tightly packed cells that separate the blood vessels in the brain from the neurons themselves. Only very small molecules can squeeze through this barrier, so anything larger, such as a protein, has to be introduced directly into the brain tissue itself, not injected via the blood stream. So some kind of device would have to be surgically implanted through the skull and into the brain. But beyond the resource and delivery problems involved, my own view is that we should not necessarily assume that agents that promote the survival of young neurons will automatically have the effect of arresting cell death in the aged brain. If aberrant activation of developmental mechanisms can indeed prove toxic in mature systems, then the effects of trophic agents administered as therapy to patients with degenerative disorders might be the very opposite of what is actually required.

Thanks to the wonders of IT, we are discovering all sorts of new ways of manipulating and using our brains. Already we take the CDROM and the internet for granted. Despite their incredibly short history, we have quickly come to depend on them as fast and convenient sources of information. All over the world, vast amounts of information are becoming accessible through the internet. The complete, unabridged content of the *Encyclopaedia Britannica*, for example, is now free on line; few schools in poor countries could afford the full set of books, but internet access is getting cheaper every year. By 2020, it is estimated that a staggering 75% of the world's population will have internet access. Before long, nearly every newspaper, journal and lecture will be instantly available on line, which can only be a good thing.

But think back to the incredible plasticity of the brain that we witnessed in chapter three. For the last 100,000 years, even though we have shared similar environments and experiences with our families, they have not occupied the same time and space points. We have had highly unique experiences. If each experience does indeed leave its mark by shifting slightly the connectivity of the brain, then each subsequent experience will, in turn, be interpreted differently. As we saw earlier our lives will be spent evolving a highly individual brain as a result of a unique sequence of highly specific experiences. Will the internet make us more or less standardized?

On the one hand, it is undeniable that the internet gives people a wider choice of sources of information and there is no doubt that more of us will have more access to information than would ever have been thought possible even a decade ago. But information is not knowledge. Knowledge surely comes only with wisdom – the placing of events, people and feelings into an increasingly complex and idiosyncratic context. There is no reason to think that the internet, by bombarding us with masses of facts, will increase our wisdom. Yet perhaps we are heading towards a future where simply amassing facts becomes more important than developing a personal interpretation, an individual understanding. If the emphasis shifts from developing knowledge to amassing information, it could change the sort of people we become and the sort of society we live in.

◄ Virtual tourism. The visitor to this Hemispherium is able to see the interior of the Guisborough Priory, North Yorkshire, as it would have appeared in the 12th century.

In addition, the pictures and sounds of the multimedia experience give our imagination no exercise, unlike a low-tech book. When we learn to read as children, pictures first dominate, but these are gradually replaced by text as our reading skills improve. Eventually, we acquire the awesome ability to conjure up whole scenes in our imagination alone at the mere sight of words on a page – we can imagine ourselves in the middle of a Napoleonic sea battle or caught in a web of espionage. If the internet makes books obsolete, will we become sensation-junkies hooked on the thrill of the moment, with a 3-second attention span and an underdeveloped imagination? The issue is whether the internet and multimedia threaten to undermine the intellectual skills that underpin society. If our brains are not trained from early childhood in the art of concentration and imagination, surely we will lose out.

Virtual reality (VR) leaves even less to the imagination than the bells and whistles of CDROMs and the internet. The VR headset completely encloses the head, providing a total audiovisual experience and blocking out any view of reality. In one futuristic scenario, it might be possible to rent a virtual day of life, just as one rents videos today. Imagine what might happen to the connectivity of our brains if we were subjected continuously to VR from birth, with a computer-generated mother cooing over us, giving each child an identical experience of life. What sort of homogenized society might result? Of course, this is an extreme example, but it serves to remind us of the importance of nurture in shaping our development as individuals.

Yet we can envisage even more phantasmagorical possibilities than these. Already there have been predictions that we will one day be able to implant silicon chips into our brains, establishing a silicon–carbon interface. And some have predicted the reverse, speculating that it will become possible to download our memories, even our whole selves, onto computers. Let's unpack each of these ideas in turn.

Kevin Warwick, Professor of Cybernetics at Reading University, is planning to have an implant put in his arm to monitor his nervous system. The implant will be able to receive signals from a computer, which, as a result, will be able to influence Kevin's emotions. It is the start, Kevin thinks, of a blurring in the dividing line between man and machine. Already, electrical engineers Ian Pearson, Chris Winter and Peter Cochrane of British Telecom's laboratories are predicting the rise of *Homo cyberneticus*. They envisage a computer interfacing directly with a human, generating artificial senses and reading a person's thoughts. One way this might be achieved, they speculate, is to use microscopic probes that connect to synapses. But we know that the brain just doesn't work like that. As we saw in the previous chapter, consciousness involves millions of neurons forming transient assemblies; tinkering with individual synapses would have little effect.

Although it just might be possible, in the misty future, to achieve the technical feat of implanting a silicon chip into someone's brain, what would be the point? Presumably the chip would supply some sort of data, equivalent to memories. As we saw in chapter five, there are two main types of memory: memory for facts and memory for events. In the case of memory for events, even the simplest scene is predicated on a whole range of other memories, facts and emotions. To implant a memory of an event would require implanting all the associations that went with it. For example, imagine trying to implant someone's memory of a trip to the seaside with their grandmother. The memory would need to include that person's feelings about their grandmother, the history of the relationship, how much they liked the seaside, how tired they were

▼ A door opens automatically for Kevin Warwick: the result of a radio signal sent by a silicon transponder implanted in his arm.

that day, and so on *ad infinitum*. To be sure that every related association and nuance was there, one would need to implant the entire mind.

What about restricting our ambitions to implanting simple factual memories, and so avoiding the time-consuming process of learning? But we can already access facts fast and furious from the internet, so why import the storage of those cumbersome facts into the physical space between the ears? In any event, the great power of the human brain is not in storing facts but in having ideas. And ideas – at least original ones in which a previously unconnected link is seen between disparate facts – would never be generated by some slab of silicon data squashed in among the neurons.

So, what about the reverse – downloading the brain onto a computer and achieving immortality in silicon? Although our biological selves might die, our essence, our minds, would be available on CD. Again, this proposition is based on the confusing assumption that a memory can be isolated and downloaded. Yet if, as just mentioned, every memory exists in a nested grouping of other memories, which, in turn, rely on the integrated operations within the whole body, then no memory could be accurately downloaded unless the whole body was too. Such a scenario, I think, is completely impossible.

If we cannot achieve immortality in silicon, can we preserve the mind indefinitely some other way? The thinking is that future generations will have the medical technology to revive the frozen brain, cure whatever ailment led to death, and transplant the brain into a new body. But the chances of such a scheme succeeding are slim indeed. Muscles might be readily frozen and thawed, like a Sunday joint, but brains are not quite so resilient. A frozen brain, once thawed, is not so much like a leg of lamb as a soggy strawberry.

But there may be other ways of cheating time. We know that neurons can survive being cooled not to freezing point but to a more modest 4°C (39°F), the temperature of a fridge. Action potentials stop completely, but normal activity resumes when the cells are warmed up again. Now, imagine if a technique were perfected for cooling the brain and body so that all metabolism, and with it all brain activity, could be put on hold – not unlike the coma induced by the apothecary's drug in *Romeo and Juliet*, which fooled Romeo into thinking his great love was dead. What kinds of minds might we have if we lived intermittently in this way? One could not assume that people the same age had lived in the same time or shared the same knowledge base. The individual would feel remote and isolated, cut off from the rest of society. Such is the stuff of science fiction. But as we know only too well, science fiction sometimes becomes reality.

▲ The Alcor cryonics laboratory in the USA, where clients are frozen in liquid nitrogen and their bodies stored inside steel cylinders for resurrection at some future date. Since 1967 around 30 people in America have been frozen in this way, in the hope of cheating death.

The ultimate sci-fi fantasy, however, would be to discover how the brain generates consciousness. We saw in the previous chapter how we might be coming close to a correlation between states of the brain and certain states of consciousness. But it is a very different matter to understand the causal link, how events in the brain actually produce a subjective sensation. In order to understand that, it would be necessary to experience first-hand the dawning of a moment of consciousness, while at the same time objectively observing what was going on.

Imagine, for instance, an experiment in which I observe consciousness dawning in someone. If I had every conceivable scanning device, I could monitor brain wave patterns, blood hormones, changes in arousal, transient neuron assemblies, and so on. But the only way I could tell when consciousness was really beginning would be to tap into that person's subjective world. The only way to do that would be to somehow jump into their body, which is as impossible as it is absurd. So, rather than trying to observe subjective consciousness in someone else, I could try to observe the first-hand experience dawning in myself. But this involves a paradox. I would need to be unconscious to begin with, so I would be unable to objectively watch what was happening.

But let's indulge our imaginations and pretend we've solved the problem of consciousness. What would life be like then? If we really understood how the brain generates consciousness, then I could not only read your mind indirectly, as a scientist observing your brain scans, but first-hand as well; I would feel directly what it was like to be you, and if I could do that then you and I would essentially be the same person. We would cease to exist as individuals.

The human brain is an incredible organ. Somehow it generates emotion, language, memories and consciousness. It gives us the power of reason, creativity, intuition. It is the only biological organ able to scrutinize itself and ponder on its inner workings – yet, despite its best efforts, those inner workings remain shrouded in mystery.

Collectively, human brains are transforming our lives beyond all recognition. We have it in our power to create a world that celebrates our individuality and the diversity of human nature, or to condemn ourselves to a living hell of uniformity and greater standardization both of nature and nurture. Even if free will is a neuronal sleight of hand – the choice really is our own.

FURTHER READING

Allman, J., *Evolving Brains*, Scientific Amercan Library, New York, 1998

Barondes, S., *Molecules and Mental Illness*, Scientific American Library, New York, 1993

Bear, M., Connors, B., and Paradiso, M., *Neuroscience: Exploring the Brain*, Lippincott Williams & Wilkins, Baltimore, 1996

Bradshaw, J. L., *Human Evolution: A Neuropsychological Prospective*, Psychology Press, Hove, 1997

Calvin, N., *How Brains Think: Evolving Intelligence Then and Now*, Weidenfeld & Nicolson, London, 1996

Carte, R., *Mapping the Mind*, Weidenfeld & Nicolson, London, 1998

Damasio, A. R., *The Feeling of What Happens*, Arrow, London, 2000

Deacon, T. W., *The Symbolic Species: The Co-Evolution of Language and the Human Brain*, Allen Lane, The Penguin Press, London and W.H.Norton, New York, 1997

Dennett, D., *Kinds of Minds: Towards an Understanding of Consciousness*, Weidenfeld & Nicolson, London, 1996

Goldstein, J. A., *Addiction: From Biology to Drug Policy*, W.H.Freeman, New York, 1994

Greenfield, S. A., *The Human Brain: A Guided Tour*, Weidenfeld & Nicolson, London 1997; paperback: Phoenix Press, 1997

Greenfield, S. A., *The Private Life of the Brain*, Penguin, London, 2000

Hobson, J., *Consciousness*, Scientific American Library, New York, 1999

LeDoux, J., *The Emotional Brain*, Weidenfeld & Nicolson, London, 1998

McCrone, *Going Inside*, Faber & Faber, London, 1999

Mithen, S., *The Prehistory of the Mind: The Cognitive Origins of Art, Religion and Science*, Thames & Hudson, London, 1996

Pert, C., *Molecules of Emotion*, Scribner, New York, 1997

Ramachandran, V. S., and Blakeslee, S., *Phantoms in the Brain: Human Nature and the Architecture of the Mind*, Fourth Estate, London, 1998

Rose, S., *Lifelines: Biology, Freedom, Determinism*, Penguin, London, 1997

Rose, S. (Ed), *From Brains to Consciousness? Essays on the New Sciences of the Mind*, The Penguin Press, London, 1998

Samuel, D., *Memory: How we use it, lose and can improve it*, Weidenfeld & Nicolson, London, 1999

Schneider, S. H., *Drugs and the Brain*, Scientific American Library, New York, 1996

Wall, P., *Pain: The Science of Suffering*, Weidenfeld & Nicolson, London, 1999

Whybrow, P., *A Mood Apart: A Thinker's Guide to Emotion and Its Disorder*, Picador, London, 1997

Zeki, S., *A Vision of the Brain*, Blackwell Scientific Publications, Oxford, 1993

PICTURE CREDITS

GLOSSARY

Acetylcholine   A neurotransmitter working in many parts of the brain and body, including neurons of the basal forebrain (the part of the brain damaged in Alzheimer's disease). Acetylcholine plays a key role in triggering arousal. Nicotine acts as an imposter for acetylcholine, which is why cigarettes boost arousal and help concentration.

Action potential   The technical name for the electrical signal that buzzes through a neuron.

Agnosia   Loss of recognition, despite normal working of the senses. People with auditory agnosia, for instance, can hear but cannot understand speech, and people with object agnosia can no longer identify objects.

Algorithm   A sequence of mathematical steps followed to reach a solution to a particular problem. A simple algorithm can be used to convert temperature in Fahrenheit to Celsius, for instance. Computer programs are essentially algorithms, albeit very complex ones.

Alzheimer's disease   The most common form of senile dementia, caused by death of neurons in the basal forebrain. Symptoms include forgetfulness, disorientation and confusion.

Amnesia   Loss of memory.

Amygdala   A deep-seated, almond-shaped structure near the front of the brain. The amygdala plays a role in emotion. Damage to it can cause Kluver–Bucy syndrome, in which a person becomes 'hypersexual' and makes indiscriminate sexual advances towards inanimate objects, such as items of furniture.

Anteretrograde amnesia   The inability to remember things that happen after the event that caused amnesia.

Arousal   When we feel alert or excited, we are said to be in a state of high arousal. Arousal is low when we are relaxed or sleeping.

Artificial intelligence   The ability of a computer or robot to perform tasks requiring some of the faculties characteristic of humans, such as learning, reasoning and language.

Axon   The output fibre that carries an electrical signal from a neuron to a neighbouring cell. Neurons have only one axon each. They are often much longer than the neuron's input fibres (dendrites).

Basal forebrain   The part of the brain primarily damaged in Alzheimer's disease. The basal forebrain is in the lower front part of the brain and sends widespread connections to other parts of the brain, such as the cortex.

Basal ganglia   A series of large clusters of neurons deep inside the brain. The basal ganglia are interconnected and have many complex connections with the rest of the brain and nervous system. One of these clusters is the substantia nigra, the part of the brain damaged in Parkinson's disease.

Blindsight   A condition in which a person can point to something that they cannot consciously 'see'. People with blindsight usually claim their responses are pure guesswork.

Blood–brain barrier   A barrier between the bloodstream and brain cells formed by a layer of special cells surrounding blood vessels in the brain. The blood–brain barrier prevents certain chemicals and drugs (particularly those with large molecules) from entering the brain.

Brain imaging   See brain scan.

Brain scan   A picture of the inside of the brain produced without the need for surgery (non-invasive). There are a number of different brain-scanning techniques, ranging from a type of X-ray (CAT scan), through to those highlighting errors of large oxygen and glucose consumption (PET, fMRI) to MEG which monitors electrical fields generated when action potentials are fired.

Brain wave   A distinctive pattern of electrical activity in the brain, revealed by a printout from an EEG machine. During dreamless sleep, a person's brain waves are slow and regular. During dreaming or wakefulness, they become faster and more erratic.

Brainstem   The core part of the brain, where it merges into the top of the spinal cord.

Broca's aphasia   A speech impairment in which the patient can understand everything said to them but is unable to articulate.

Broca's area A part of the left frontal cortex near the temple. Damage to Broca's area causes a speech impairment known as Broca's aphasia.

Capgras' syndrome A rare psychological disorder in which the patient believes that close friends and relatives have been replaced by sinister imposters.

Cell body The control centre of a neuron. The cell body integrates inputs from all the dendrites. If the resulting change in voltage is sufficiently depolarized (potential difference reduced), then an action potential is generated and propagated down the axon.

Cerebellum A cauliflower-shaped structure at the back of the brain. The cerebellum is involved in well-practised physical skills requiring coordinated input from the senses, such as playing the piano or driving a car.

Cerebral hemisphere See hemisphere.

Cerebrospinal fluid A watery liquid flowing around the brain and spinal cord, this is taken in samples of lumbar puncture.

Cerebrum The largest and uppermost part of the human brain. The cerebrum is divided into two halves (left and right) called cerebral hemispheres, each of which is divided into four main lobes. The outer part of the cerebrum is known as the cortex and is deeply folded. The cerebrum is much larger in humans than in other animals.

Consciousness The first-hand experience of being alive, that is lost during sleep or anaesthesia. Dreaming is arguably a form of consciousness.

Cortex (cerebral cortex) The deeply folded outer layer of the largest and uppermost part of the brain (the cerebrum). Humans have a disproportionately large and wrinkled cortex.

Cortisol (hydrocortisone) A hormone that helps the body cope with stress by boosting blood sugar levels. Cortisol also has a powerful anti-inflammatory effect. Drugs with a similar action are called corticosteroids.

Delayed-response test A test that assesses working memory. For instance, a monkey might be trained to take food from one of two containers. It is then shown the food being transferred to the second container and allowed to make a choice. Animals with impaired working memory continue to reach for the first container, but normal animals reach for the second.

Dementia Deterioration in mental ability. The most common form of dementia is Alzheimer's disease, which affects older people. Dementia can also be caused by head injuries, brain tumours, alcoholism and a variety of diseases.

Dendrites The input fibres that carry electrical signals into a neuron from connected cells. A single neuron may have up to 100,000 inputs onto its dendrites.

Dopamine A neurotransmitter that, among many other functions, plays an important role in movement. Damage to one of the dopamine fountainheads (the substantia nigra) leads to Parkinson's disease, which results in rigidity, tremors and inability to generate movement.

EEG (electroencephalogram) The record from an EEG machine (electroencephalograph), showing the pattern of electrical activity (brain waves) in a person's brain. The electrical activity is detected by a number of electrodes placed on the person's skull. EEG is used to diagnose brain disease and is also useful in research.

Electroconvulsive therapy (electric shock treatment) A treatment for intractable cases of clinical depression. The patient is anaesthetized and then given a controlled electric shock across the temples to induce brain seizures. Electroconvulsive therapy can sometimes be very effective, but scientists do not understand how or why it works.

Epigenetic factor Something other than a gene that affects how a cell develops. Developing neurons are not programmed by their genes – the way they develop and the cells with which they form connections, depend to a large extent on epigenetic factors in their environment. An example is nerve growth factor, a chemical that stimulates growth of axons and dendrites.

Epilepsy A condition characterized by recurrent, transient seizures resulting from abnormal electrical activity in the brain. Epileptic seizures are very variable and can take the form of major physical convulsions and/or momentary blackouts. There are many different causes of epilepsy.

Episodic memory Memory of events and experiences anchored in a specific location and time.

fMRI scanning A brain-scanning technique that reveals which areas of the brain are working hardest by detecting faint radio signals from the oxygen-carrying chemical in blood (haemoglobin). fMRI stands for functional magnetic resonance imaging.

Frontal cortex The outer layer (cortex) of the frontal lobe.

Frontal lobe The front lobe of each half (hemisphere) of the main part of the brain (the cerebrum).

GABA (gamma-aminobutyric acid) A neurotransmitter that works by inhibiting the action of the target cell. Implicated in anxiety and epilepsy. Some neuroscientists think it might operate in up to 30% of the brain's synapses.

Gene therapy An experimental treatment that aims to cure genetic disease by supplying the patient with a working copy of a faulty gene.

Genetic screening A test for a faulty gene that could cause disease. Parents sometimes undergo genetic screening to assess their risk of passing on a genetic disease to children.

Glial cell A type of brain cell that provides a supportive role to neurons, repairing damage and so on. Glial cells do not transmit electrical signals but they are vital to the healthy working of the brain.

Golgi stain A procedure that makes one in ten neurons visible under the microscope.

Grey matter Brain tissue that contains many neuron cell bodies and so looks slightly darker in colour. White matter is brain tissue made up mainly of axons, with few cell bodies.

Hemisphere One of the two halves (left and right) into which the largest and uppermost part of the human brain (the cerebrum) is divided.

Hippocampus A part of the brain deep below the temporal cortex. Damage to the hippocampus causes certain types of memory loss.

Hormone A chemical that is produced by the body and circulates through the bloodstream, with widespread effects. Hormones are important in body housekeeping and help to regulate hunger, thirst, sex, growth and the immune system, among other things. Testosterone, oestrogen and adrenaline are hormones.

Human Genome Project A major international scientific project that aims to map every gene in human DNA.

Huntington's chorea A disease characterized by wild, involuntary flinging movements of the limbs (hence the name chorea, from the Greek word for dance). Huntington's chorea is caused by damage to a part of the brain called the striatum.

Hypothalamus A structure like a flat grape that lies almost on the floor of the brain, just in front of the brainstem and behind the eyes. The hypothalamus receives inputs from throughout the body and plays an important controlling role in the body's hormone system, affecting temperature, hunger, thirst, growth, sex and the immune system.

Inferotemporal Important in visual recognition, i.e. an area where visual systems and memory systems mesh.

Ion An atom that is positively or negatively charged. Neurons use ions to transmit electrical signals.

Ion channel A gateway in the membrane of a neuron that lets ions (electrically charged atoms) in or out of the cell. Ion channels open and close in swift succession as an electrical signal passes through a neuron. Opening of these channels is controlled by a change in voltage or by the complexing of a transmitter to its receptor

Large-cell system (System A) One of two parallel networks of neurons involved in processing vision. The large-cell system deals with perception of movement.

Larynx (voice box) A chamber at the top of the windpipe containing flaps of tissue that vibrate as air passes between them, so generating sound.

Lateral geniculate nucleus A part of the thalamus. This particular region processes signals coming from the eyes and then forwards the signals to the part of the cortex concerned with vision. There are two lateral geniculate nuclei, one on each side of the brain. Each receives inputs from both eyes.

L-DOPA (levodopa) A drug used to

treat people with Parkinson's disease. L-DOPA is converted into the neurotransmitter dopamine in the brain, helping to replenish the abnormally low level of dopamine in Parkinson's disease.

Lesion   A general term for an area of tissue damage anywhere in the body, which may be experimentally induced or spontaneously occurring.

Limbic system   A conglomeration of brain regions around the brainstem. The limbic system includes the hippocampus, amygdala, hypothalamus and other regions.

Lobe   One of the four major folds in each of the two cerebral hemispheres (the two halves of the cerebrum). The four main lobes are the parietal lobe, the temporal lobe, the frontal lobe and the occipital lobe. These are named after the overlying bones of the skull.

Lobotomy (leucotomy)   A surgical procedure in which cuts are made in the prefrontal cortex in order to change a patient's personality and make them less aggressive.

Long-term potentiation (LTP)   The development of a stronger response in a neuron after repeated high-frequency priming stimulation. LTP might be one of the mechanisms involved in learning and memory.

Medial temporal lobe   A large lobe of the brain near the ear. It includes the temporal cortex, the hippocampus and other structures. Damage to the medial temporal lobe can cause impairments of memory. This area shows clear changes of degeneration during Alzheimer's disease compared to age-matched controls.

MEG scanning   An incredibly sensitive brain-scanning technique that picks up magnetic fields generated by electrical blips in the surface of the brain (the cortex). MEG (magnetoencephalography) scanning is very fast and so can reveal the cortex working at full tilt, unlike slower techniques such as fMRI and PET scanning.

Membrane   The outer wall of a cell with a composition a little like an oily sandwich, i.e. consisting of two layers with a fatty inside.

Modulation   An effect that neurotransmitters sometimes have on target cells. Instead of triggering an electrical signal in the target cell, modulation alters the way the target cell responds to a second neurotransmitter, for instance by making it more sensitive than it would otherwise have been.

Monochromatism   An extreme form of colour blindness in which the world is seen only in shades of beige or grey.

Neural net   An electronic circuit or computer program modelled on a biological nervous system. Neural nets share some features in the way they adapt with those of living neuron networks.

Neurodegeneration   The gradual and inexorable death of key groups of brain cells. Neurodegeneration is the underlying cause of Alzheimer's disease and Parkinson's disease.

Neuron   A nerve cell. Neurons can generate electrical signals and pass these on to connected cells. As such, they are the most important building blocks in the brain and the nervous system. The term neuron is often used to mean brain cell, although neurons are found throughout the body.

Neuroscience   The scientific study of the brain and nervous system that embraces a variety of disciplines, including anatomy, physiology, biochemistry and computer science. This discipline tends to focus on the more 'bottom up' of cell function, i.e. works at the level of cells and cell networks.

Neurotransmitter   A special chemical used by brain cells to pass signals from one cell to the next. There are a wide range of these chemical messengers, each with their own particular distribution in the brain

Nitric oxide   A simple chemical known to chemists as a colourless and toxic gas. In the 1980s it was discovered that nitric oxide plays various roles in the human body, helping to trigger erections and arm the immune system, for instance. It can also act as a neurotransmitter.

Noradrenaline (norepinephrine)   A neurotransmitter, one of whose effects is to put the body on red alert during frightening, stressful or exciting situations. Noradrenaline also works as a hormone, circulating in the bloodstream with its close chemical cousin adrenaline (epinephrine). Drugs such as cocaine and amphetamine, as well as some antidepressants, work by boosting the level of noradrenaline.

Nucleus The part of a cell where genes are stored, encoded in the chemical DNA (deoxyribonucleic acid). This term is also used, somewhat confusingly, to define a brain region such as lateral geniculate nucleus.

Oscilloscope A device that displays changes in an electrical signal as a line flickering up and down across a screen. The electrocardiogram (ECG) machine used in some hospitals to trace over a sub-second timescale is a type of oscilloscope that monitors electrical activity in the heart.

Oxytocin A hormone, produced by the pituitary gland, that causes contractions during childbirth and stimulates the flow of milk during breast-feeding.

Parietal cortex The outer layer (cortex) of the parietal lobe.

Parietal lobe One of the four main lobes that make up each half (hemisphere) of the main part of the brain (the cerebrum). The two parietal lobes are at the top of the back of the head, one in the left hemisphere and one in the right.

Parkinson's disease A disease caused by death of brain cells in the substantia nigra. Patients lose the ability for voluntary movement. In addition their muscles become rigid, and their hands display tremor.

Peptides Very common biological compounds that can act as neurotransmitters, with which they are often co-stored inside neurons. Unlike conventional neurotransmitters, they can be released outside synapses to exert a wide sphere of influence on many target cells at once.

PET scanning A brain-scanning technique that reveals which areas of the brain are working hardest by detecting radioactive oxygen or glucose injected into a person's body prior to scanning. The scan, displayed on a computer screen, is a rather blurry colour pattern, with active regions highlighted in colour. PET stands for positron emission tomography.

Phrenology A 19th-century doctrine that attempted to relate aspects of personality to the shape of a person's skull.

Pick's disease A type of dementia caused by disintegration of tissue in the temporal cortex. The usual symptoms

are loss of memory for facts, but some patients also develop surprising new talents, such as an interest in painting.

Pituitary gland A pea-sized blob of tissue that dangles on a stalk from the underside of the brain and fits snugly into a hollow within the skull. The pituitary gland produces important hormones affecting growth, sexual development and other body processes.

Prefrontal cortex The outer layer (cortex) of the front lobes of the brain, behind the forehead. Compared to other animals humans have a colossal prefrontal cortex. As a result, this part of the brain is often seen as the site of higher functions, such as language, intelligence and consciousness.

Prosopagnosia A condition in which a person is unable to recognize faces, despite having normal vision. Prosopagnosia is caused by damage to part of the inferotemporal cortex.

Psychoanalysis A form of psychiatric treatment, based originally on the theories of Sigmund Freud, that aims to treat emotional or behavioural problems by helping the patient understand their subconscious drives and conflicts. Psychoanalysis has changed since Freud's time. Today, analysts attempt to relate a patient's problems to emotional traumas experienced in childhood.

Receptor When a neurotransmitter crosses a synapse it binds to a receptor molecule embedded in the membrane of the target cell. This chemical 'handshake' causes ion channels to open in the target cell, triggering an electrical signal.

REM sleep (rapid eye movement) A stage of sleep when the eyes move quickly back and forth. People woken during REM sleep usually report that they have been dreaming.

Retrograde amnesia An inability to remember things that happened before the event that caused amnesia.

Schizophrenia A severe disorder of thought, where the patient may hear imaginary voices, suffer from paranoia, hallucinations, depression and illogical thoughts. Contrary to popular belief, the disease does not cause split personality. About 1 in 100 people worldwide are affected by schizophrenia, and about 1 in 10 people with the disease commit suicide. It is frequently associated with a functional excess of the transmitter dopamine.

Semantic memory   Memory of isolated facts without reference to a specific location or time (see also episodic memory).

Serotonin   A neurotransmitter involved in many processes, including pain, sleep and mood. The drugs Prozac and ecstasy work by boosting serotonin's effects.

Small-cell system (System B)   One of two parallel networks of neurons involved in processing vision. The small-cell system deals with perception of colour and form.

Source amnesia   Inability to recall the context of a memory, such as the time or location of an event.

Spinal cord   The downward extension of the brain that runs through the spine. The spinal cord connects fibres coursing in the opposite direction from sensory receptors in the skin. Severe damage to the spinal cord causes paralysis.

Striatum   One of the parts of the basal ganglia, in the heart of the brain. Damage to the striatum causes the disease Huntington's chorea, in which patients make wild, involuntary flinging movements of the limbs.

Stroke   Damage to part of the brain due to a blockage or leak in one of the brain's blood vessels. As a result a group of brain cells become starved of oxygen and therefore die within a few minutes.

Subconscious   A mental process that takes place without our awareness while we are awake, for example while driving, but is described as subconscious.

Substance P   A peptide neurotransmitter that plays a role in pain and possibly stress. Research into substance P may lead to the development of new drugs to treat depression.

Substantia nigra   Two pigmented moustache-shaped bands of cells (grey matter) in the heart of the brain. Damage to the substantia nigra causes Parkinson's disease.

Synaesthesia   A condition in which the senses cross over, making it possible to smell colours, taste sounds.

Synapse   The gap between two adjacent neurons. Chemicals called neurotransmitters must cross a synapse to carry an electrical signal from one neuron to the next.

Temporal cortex   The outer layer (cortex) of the temporal lobe.

Temporal lobe   One of the four main lobes that make up each half (hemisphere) of the main part of the brain (the cerebrum). The two temporal lobes are behind the ear and temple, one in the left hemisphere and another in the right.

Thalamus   One of a pair of egg-shaped structures in the heart of the brain, above the brainstem, containing a variety of brain nuclei, such as the lateral geniculate nucleus. In general the two thalami are involved in processing different senses, with each nucleus relaying a particular modality to a respective part of primary cortex. In addition, other areas of the thalamus have been implicated more generally in arousal and consciousness.

Turing test   A test intended to determine whether a computer can think. A person asks questions of a computer and a person without knowing which is which and attempts to identify them from the answers. If the computer cannot be identified, it has passed the Turing test. So far, no computer has passed, although a human being has failed!

Visual cortex   The part of the cortex concerned with vision.

Wernicke's aphasia   A speech impairment in which the patient can talk fluently but uses circumlocutions and inappropriate words. In extreme cases a patient may emit a fluent stream of meaningless nonsense, although this may sound intelligent from a distance. People with Wernicke's aphasia also have difficulty understanding what is said to them.

Wernicke's area   A part of the left temporal cortex. Damage to Wernicke's area causes a speech impairment known as Wernicke's aphasia.

White matter Brain tissue consisting mainly of the axons of neurons, with relatively few neuron cell bodies. Grey matter, which is darker, contains many more cell bodies.

Working memory   The type of memory responsible for keeping in mind information relevant to the task in hand, e.g., working memory enables you to remember where all the pieces are during a game of chess.

INDEX